How Things Are Made

How Things Are Made

How Things Are Made

A Journey Through the Hidden World of Manufacturing

Tim Minshall

ecco
An Imprint of HarperCollins*Publishers*

HarperCollins books may be purchased for educational, business, or sales promotional use. For information, please email the Special Markets Department at SPsales@harpercollins.com.

Originally published as *Your Life Is Manufactured: How We Make Things, Why It Matters and How We Can Do It Better* in the United Kingdom in 2025 by Faber & Faber Limited.

An extract from "Home Computer Terminal," *Tomorrow's World* (1967), is reprinted by kind permission of the BBC.

FIRST U.S. EDITION

Library of Congress Cataloging-in-Publication Data has been applied for.

ISBN 978-0-06-343465-3

25 26 27 28 29 LBC 5 4 3 2 1

To my family, and to all who are helping
to manufacture a better world.

Contents

Prologue

Things were not going to plan. There was no way this was going to end without someone getting hurt. '*Please*,' I implored, my voice barely audible over the shouting, '*there's enough for everyone! Be careful, some of these things are sharp! One at a time!*' It was no good. The laptop keyboard seemed to have exploded, wires hung limply from a gaping hole at the back of the PC, and Mr Incredible was clearly going to need medical help.

⬡

A week earlier, I had been sitting in my office when I received a call from a teacher at a local primary school. She wanted someone to give a talk as part of her school's 'science week'. I tried to explain that I wasn't a scientist but rather a manufacturing engineering academic. This was met with: 'Sure. Sounds great. There'll be around sixty nine-to-ten-year-olds, and the session is forty-five minutes. Monday next week at 9.30 OK for you? Turn up a few minutes before and we'll show you to the room. Thanks a lot!' *Click*.

I put down my phone and scratched my cheek. What on earth was I going to talk about? As this talk was a whole week away, I wasn't overly concerned. Six days later I sprang into action. I got up early on Sunday morning to give myself an uninterrupted block of time. Plucking a nice fresh sheet of A4 paper from the printer and a pleasingly sharp pencil from the cracked mug on my desk, I sat down and got straight to work.

The empty page stared back at me. I nudged it so that it aligned neatly with the edge of the desk. I started doodling some pictures of factories and ships. Then, I carefully wrote the words 'How does stuff get made and delivered to you?' in the middle of the page. I looked at this for a few moments before, with my brow furrowed and the tip of my tongue poking out of the corner of my mouth, slowly drawing a neat cloud around the question. I underlined each word. Twice. Chewing on the end of the pencil, I tried to suppress a queasy, rising sense of panic.

How could I possibly make manufacturing interesting for a room full of pre-teens?

Then, inspiration struck. I threw down the pencil and strode out to our shed. There, sitting on the floor covered in cobwebs, in the dark amongst bits of old bikes, were some long-dead computers. I hauled these onto a table, grabbed a screwdriver and started to take them apart. As I pulled out circuit boards, disk drives, desiccated spiders, bundles of wiring and stacks of connectors, I came across what I had hoped to find. On neat little labels or directly printed in wobbly dots on various cables and components were the words 'Made in Malaysia', 'Product of Mexico', 'Assembled in the USA', very occasionally 'Made in the UK' and most frequently 'Made in China'.

I returned to my desk with a plan.

We all live in a manufactured world

Despite the pervasiveness of the online, the virtual, the cloud-based and the app-enabled, we cannot survive without real, physical, manufactured things.* Just look around where you are now. Everything you can see – other than plants, rocks, people and other

* And that includes the online, the virtual, the cloud-based and the app-enabled – all of those are made possible by a network of computers, cables and satellites. Someone has to design, make and maintain that whole massive system-of-systems.

animals – has been made by someone and delivered to where it is needed. Unless you are currently floating naked through space, you are right now in immediate contact with multiple manufactured products. Throughout every day of your life you will be wearing, consuming, being transported or sheltered by, communicating through or being restored to health by manufactured products.

Yet the processes by which these items appear in our lives are, to most of us, largely invisible.

And that has some worrying consequences.

It hasn't always been this way. Go back three centuries and you'd be much more likely to have close, even personal connections with those who made the things you needed – from tailors to potters, butchers to bakers, blacksmiths to carpenters. I am not naively painting some rose-tinted picture of pre-industrial revolution life, but a shorter distance between production and consumption had some advantages. Local production made visible to the immediate community any waste or pollution being generated in the process; we would know about poor working conditions because the people enduring them would be from our village or town; and we would see that the money we bought things with supported our neighbours' families.

Over time, thanks to industrial revolutions, the rise of consumerism and globalisation, the bonds between those who make and those who buy have been stretched thin to the point of near invisibility. I have absolutely no connection with the people involved in the making of the laptop on which I am typing these words – I clicked a few buttons on a website four years ago and a couple of days later this miracle of engineering arrived in a box at my front door. I have similarly remote connections with those who made my pens, chair and coffee mug, and it's the same for those who made the clothes I am wearing, the house in which I am sheltered from today's showers, the car and bicycles parked outside, the food in my fridge . . . you get the picture. Manufacturing has become like the sewage system: essential for our lives, yet out of mind until things go wrong.

And recently, a lot of things have been going wrong.

When manufacturing goes bad

We all need things, we all want things, so things get made. We get the products we require or desire; manufacturers get the income needed for their businesses to thrive, to provide jobs for their workers and returns for their owners. (Or, reversing the equation, manufacturers sometimes make things and then create desire through advertising.) The long-overdue realisation of our need to focus on planetary sustainability, however, is changing everything. Manufacturers know that splashing some green over their logos or sticking a picture of a tree or a dolphin on their packaging is not going to cut it. We as consumers and those who produce for us now (in most cases) have shared values focused on behaving in a way less likely to damage or destroy our world.

This has been a remarkable and rapid change, one that can be seen in the rocketing sales of electric vehicles, the transformation of our attitudes to single-use plastics and the frantic competition amongst firms and nations to announce that they will be 'net zero' by 2050/2040/2030.*

This change in attitude has arisen as the connections between production, consumption and the survival of our world have become shockingly clear. In this context, some see manufacturing as 'a problem that needs fixing' (quite reasonably, given that manufacturing activities constitute one of the biggest contributors to global CO_2 emissions).[1] But manufacturing is also a key part of the solution. Manufacturing firms are going to make the

* Delete as applicable.

machines that will give us widescale access to renewable energy, ensure sustainable sources of food and deliver clean transportation. So could we make *all* our manufacturing truly sustainable? And can we, as individuals, do anything to help?

The answer to both of those questions is yes, and this book will explain how.

At its heart, manufacturing is all about making physical items (production) and moving them to where they are needed (logistics).

Everything we buy or use must be moved from where it was produced to where it will be used. My desk, and everything scattered over it – pens, Post-its, laptop, phone-resetting paperclip, desk lamp, Tintin figurine – had to be made and then physically moved, in most cases hundreds if not thousands of miles. We get to glimpse some of these movements: we see shipping containers on the backs of trucks, ubiquitous white delivery vans, occasional freight planes flying overhead and cargo ships floating on the horizon glimpsed from the shore.* While we are able to see snippets of those product journeys, we rarely appreciate their scale and complexity – and we almost never see the actual production.

It's as if there are two worlds. There is our visible 'regular world': our homes and places of work, study, retail, leisure and healthcare. Then, there is the partially visible 'manufacturing world': the mines, farms, factories, processing plants, warehouses, construction sites and power plants that feed the regular world. This second world is either all-too-clearly externally visible – it's pretty hard to hide a car factory, an Amazon warehouse, or an oil refinery – or completely out of sight owing to it being on the other side of the world. But very few of us get to see inside and understand what is going on within the innards of the 'making' part of the manufacturing world.

* For those who want to geek out, you can even see where all those planes and boats are coming from and going to using apps like www.flightradar24.com and www.marinetraffic.com.

The near-disastrous session with those primary schoolchildren on that grey Monday morning in Cambridge was one of my early exploratory attempts to bridge these two worlds.

The laptop keyboard disintegrating as it smashed into the floor marked a turning point of the school assembly.

'Stop!' The room fell silent as my host deployed her far superior crowd-control skills. 'Everyone, go back to your seats.'

I swiftly turned to look for mine before realising she wasn't talking to me. As the children shuffled and giggled their way back to their chairs, my pulse lowered below 150 and I surveyed the aftermath of the riot.

To my surprise, the result was kind of what I had hoped for, even if the process had been a bit more chaotic than intended. There, spread out on the tables at the front of the room was a big map of the world and, stacked in neat-ish piles over different countries, the components of dismembered computers, phones and toys. The biggest pile was over China, and atop the circuit boards and microchips, the Mr Incredible figurine sat, head horribly twisted, staring mutely back at me. Via stage-managed(ish) destruction, the children had been able to produce a simple, tangible model of a supply chain: the globally scattered operations of making and moving that need to be co-ordinated and connected to deliver something to you.

With this book, I aim to provide you with a deeper and broader-ranging version of that schoolroom exercise (with less screaming and fewer tellings-off from a teacher). As I set out to start writing, tragic global events made crystal-clear why we all need a better understanding of what is happening within the manufacturing world.

The surprising fragility of our manufacturing systems

Back in 2020, consumers in wealthy countries stared in disbelief at empty supermarket shelves. They were stunned by the sudden inability of online retailers to deliver whatever they wanted within twenty-four hours.

Stories about shipping container costs and labour shortages in Chinese factories that would have previously appeared only in niche publications such as the *Journal of Shipping and Trade* or buried in articles in the *Financial Times* started appearing in our regular newsfeeds. Many people were introduced to the sometimes daunting language of manufacturing for the first time. This is a world where the assembling of swaged, flanged and fettled components takes place in an acronym soup (BOM, ECN, SPC) enriched with the language of logistics, be this inside the factory (where runners, repeaters and strangers are moved by SLAM-co-ordinated AGVs) or beyond (where SKUs are packed into TEUs – using appropriate dunnage – with agreed Incoterms to avoid any demurrage).*

As pandemic-induced chaos hit the manufacturing world, commentators and politicians scrambled to understand, communicate and respond to what was going on. My colleagues, who had become resigned to never talking about their work at social events, suddenly found themselves the centre of attention at parties (well, at least at some parties). These same colleagues knew their time in the spotlight had arrived as they watched prime ministers and presidents confidently combining the words 'supply', 'chain' and 'resilience' as they read from their autocues. And cries of *What is going on!? Why can't we just make the things here?* led to awkward press briefings where many politicians struggled to explain why we couldn't instantly start or ramp up local production of things like surgical masks, rubber gloves and medical ventilators.

* Don't be alarmed. This book and associated website will provide an explanation of all the necessary vocabulary to explore the manufacturing world.

So, not only is the manufacturing world damaging our planet, but it is susceptible to damage itself, leaving us vulnerable to shortages in our times of need, as the COVID-19 pandemic and many other crises since have revealed.

So, what's it really like in that 'other world'?

This book seeks to hammer wider the fractures that recent events have made in the wall between our 'regular' and 'manufacturing' worlds. But to write this book and take you through into that other world, I had to overcome a problem. As I planned the contents, I started to realise that I had some glaring gaps in my knowledge on some specific aspects of manufacturing. This was partly the result of something that affects all of us, and it has a name: *the illusion of explanatory depth.*[2] Meaning? Most of us know a lot less about some things than we believe or would like to admit. We convince ourselves that we could explain something if asked, but if we need to describe that thing in any detail, we start to struggle.[3]

So, I went to visit a lot of factories, listened to lots of very smart people, did a lot of reading and watching and then stitched together all I had learned. Luckily, as I work at the University of Cambridge in an organisation called the Institute for Manufacturing, I had access to a wide network of people who could – and kindly did – help. These people ranged from those who work in manufacturing companies of all sizes to those who design government policies to support manufacturing, from specialists advising or investing in manufacturing firms to academics exploring the cutting edge of manufacturing technologies.

I then used another tool at my disposal that arises from my day job: our amazing students. Helping very smart students learn about manufacturing is a great way to ensure I have a good understanding of all the key aspects and how they fit together. As rightly challenging as our students are, though, they are nothing compared to the ever questioning brains of primary school children. Our institute's programme of outreach to schools in our community allowed me to test the quality of my explanations with those toughest of crowds.[4] By reading this book, you are benefiting from the repeated intellectual beatings I have received via years of working with Cambridge students and the multiple 'character-building' school assemblies and classes I have given to hundreds of children.

The structure of this book also reflects my own changing understanding of the world of manufacturing. For my first eighteen or so years, in common with almost everyone I knew growing up in a very rural corner of the UK, I barely gave a thought to how things arrived in my world. Through a very non-linear career path, manufacturing gradually became a core part of my working life. But the more I grew to understand and appreciate it, the more I became troubled by how little so many otherwise well-informed and smart people seemed to know – or frankly care – about what goes on inside this other world. And, as we've already seen, the impact of climate change and recent crises has revealed with stark clarity the dangerous consequences of not paying it due attention.

I want to help you reach that realisation much faster than I did. However, in around eighty thousand words, I know that there is no way we are going to be able to cover everything. There are so many enticing rabbit holes down which we cannot dive, so to provide an outlet for those who want to know more, there is a website to accompany this book.* That website includes suggested 'Further Readings' along with videos and other resources exploring different aspects of topics covered in the next two hundred and sixty-three pages.

* www.timminshall.com

What this book will do for you

Before you read any further, I want to make what I hope is a reassuring disclaimer: this is not going to be a preachy book. I like 'stuff', especially gadgets, cars and planes, and I love – and see huge value in – international travel. I don't really want to have to make a confusing ethical judgment every time I buy something. If I could, I would keep my click-and-deliver life. Part of me wants to believe that the woes of the manufacturing world and its impact on planetary survival are all someone else's problem, and that things are all somehow going to be fine. But they won't be unless something changes, and changes *now*. The reality is that you and I need to change what and how we choose to buy. If we understand more about what goes on in the manufacturing world and how it interacts with every aspect of our lives, we can make choices that are better for our families, our communities and our planet.

This might be slightly overselling, but by the end of this book you should have an amazing new superpower. The next time your finger or cursor hovers over the 'Buy Now' button, or when you find yourself staring at a packed supermarket shelf or wandering around a shopping mall, you will be able to appreciate what has led to each product coming into existence and being made available for you, and the cascade of events that will result from your decision to buy. And at the risk of being overdramatic and clichéd, this power really does come with great responsibility: your use of this could help slow – or even start to reverse – the damage we are causing to our natural world and improve the lives of future generations.

So, no pressure.

To help you develop this new superpower, we need to head off on a journey to seek the answers to two big questions. Why is our modern manufacturing world so fragile and so damaging to the planet? And what can be done to make it less fragile and damaging?

To do that, we need to delve into the innards of this other world, to see why we make, move and consume things the way we do. Then, we need to look at how things are changing for the better, and what you and I can do to accelerate these changes.

Join me now in taking the first step of that journey by exploring the extraordinarily convoluted tale of the making of one particular soft, strong and absorbent product.

PART ONE

How Things Work in the World of Manufacturing

1: Magic

The jaw-dropping complexity of manufacturing even the simplest thing

At 6.30 on a dark March morning back in 2020, I reached from under the covers to silence my beeping phone. Drawing it to my face, muscle memory guided my thumb to the BBC News app. Squinting at the screen, my half-awake brain struggled to decipher this headline:

> Coronavirus: What's behind the great toilet roll grab?[1]

Typing that sentence now, 'the great toilet roll grab' has a comedic tinge to it. Some news outlets took things to amusing lengths: *Australian newspaper prints extra pages to help out in toilet paper shortage.*[2] Others sneered on social media at those who rushed to stockpile toilet roll (while discreetly clicking '+1' on their own shopping orders) as the implications for daily life of COVID-19-related lockdowns became clear. Those empty shelves in the supermarket personal hygiene aisles were symbolic of a much bigger and more worrying issue.

The global pandemic of the early 2020s revealed how extraordinarily delicate the system is that connects all the companies and people that make, distribute and sell us consumer staples such as food and toiletries. When it works well, we take it for granted; when it judders or grinds to a halt, we are jolted to disbelief and outrage. Why is a system that is essential for our daily lives so fragile that it collapses at the first sign of trouble?

To understand that, 'the great toilet roll grab' gives us an informative case study. Before revealing the details of that particular pandemic-induced psychodrama, though, we first need to understand how these soft sheets of paper end up in our toilets and bathrooms in non-crisis times.

How to make toilet paper

Back in 2020, my knowledge of the process of making loo roll was rather limited. I knew that trees and some kind of rolling mill were involved, but what happened between those two points and beyond was somewhat hazy.* So, I went to find out how this whole specific manufacturing system works.

To help share what I found out, I first need you to do a little practical exercise.

If you are able to do so, please put down this book or device and go and find a roll of toilet paper. Tear off one single sheet and take a good look at it. You'll see that it's a rectangle measuring about ten by twelve centimetres, possibly with some pattern or lettering on the surface. It will have wispy tufts along the short edges, the residue of the perforations that allowed you to separate it from its neighbour with a precisely calculated amount of force. You might also notice that the paper is textured or slightly crinkled – this is a feature added during the production process to improve absorbency.

If you tease the edge of the sheet, you should be able to separate it into two, three, four or – if you are very posh – possibly five separate layers (or 'ply'). If you have managed to separate the ply, run your fingers over each surface. You might be able to detect a slight difference in the texture, partly a result of the glue used to bond the layers. This is also an indication of the different lengths of fibres stacked up within each single ply to ensure the optimum balance of strength and softness. A single ply could be a blend of

* A neat illustration of that *illusion of explanatory depth* mentioned earlier.

pulp from Scandinavian trees (good for strength) with that from South American trees (good for softness).

Decades of research by engineers and scientists have resulted in the product now partially disassembled in your hand. Armies of people have been involved in the designing, making and delivering of this most basic of products to you at the right price and quality for your particular budget and preference. In countries where toilet paper is commonly used,* the average consumer will use over one hundred rolls per year.[3] Each of those sheets has made a remarkable journey to reach your hand. I'd now like to take you to a midge-infested location that illustrates one possible starting point of that journey.

From tree to pulp†

I'm standing beside a loch in the Highlands of Scotland, feeling a bit sorry for myself. The morning had started well. I'd drunk my early morning coffee sitting on a grassy slope under blue skies, listening to tweeting birds and smelling bracken warming in the sun. Now, up in the muggy damp hills, I'm sweaty, tired and bleeding. I love Scotland, but in summer, there are two things that can distract you from enjoying the stunning landscapes: the tiny midges that irritatingly cloud around and latch on to exposed skin, and the tenacious horseflies that feast on any flesh that is stupid enough to present itself before them.

Once I paused my swatting and swearing long enough to look up, what I could see laid out before me was extraordinary.

* Around 70 per cent of the world's population use other methods. For those who'd really like to dig further into human waste (as it were), Rose George's book provides a great starting point: George, R. (2008). *The Big Necessity*. Portfolio.

† This summary assumes that virgin pulp rather than pulp from recycled paper is being used; although recycled paper is becoming increasingly common, it still represents a small minority of the loo roll available.

Surrounding all sides of the insect-infested loch, carpeting the rolling hills as far as my eyes could see, were millions of densely and darkly packed identical trees. This was not leafy, ancient woodland: this was industrial production of coniferous trees, grown as a crop to provide raw material for construction, energy and, of course, paper.

In the UK, around twenty thousand people work on planting and chopping down trees. Add in all the people involved in various things that get done to the wood once harvested – saw-milling, pulping and making panels, paper and, well, anything made of wood – and the number swells to around 140,000. Together, these people are worth around £8 billion to the UK economy. Globally, the forestry sector employs more than twenty million people directly and forty-five million indirectly. And they generate $0.5 trillion and $1.3 trillion respectively. Key point: it's a big industry.[4]

Given that trees rather obstinately refuse to grow very fast, forestry is an industry with long production cycles. From seeding to felling takes decades – between forty and 150 years, depending on the type of tree and purpose for which the wood is needed.

The tree production cycle has four phases. First, there is preparation (from seed to little sapling), then planting (a very manual process of sticking each sapling into a mound of earth) and then thinning (as saplings are planted very close together to protect them from the elements during their pre-school years, the ones that survive to teenagers are then too close to each other and need weeding out before they fully mature). Finally, once the appropriate number of decades has passed, the harvesting can begin.[5] And in the same way that sophisticated technology has been developed to harvest food crops (such as giant combine harvesters dustily churning their satisfyingly neat way through golden wheat fields), the world of forestry also has some frankly terrifying-looking machines to deploy.

Pretty much everything about 'tree harvesting' or 'forestry harvesting' machines is impressive.* Long gone are gangs of axe- or even chainsaw-wielding lumberjacks in the world of commercial forestry. Instead, think Japanese anime-style Transformers but with rather dull names such as the Komatsu 931XC-2020 or John Deere 1470G, which come with multi-six-figure price tags. They all look a bit like what you'd expect some over-caffeinated engineers to come up with when asked to build a giant robot lumberjack from spare parts lying around in a farming equipment factory. These machines have big wheels or tracks (to allow off-road traversing of steep inclines), a power unit, an operator's cab (the interior of which is filled with joysticks, screens and a ludicrous number of buttons) and a long, articulated arm, at the end of which dangles a slightly mad collection of toothed wheels, chainsaws, blades and sensors. The operation of these highly analogue machines is enabled by digital wizardry that allows the operator to plan and manage the whole felling process with military precision.

Before the first tree is touched, the operator plans each phase of the chopping, marking out areas for storing logs and ensuring that trees not yet ready for the brutal embrace of the 931XC-2020 or 1470G remain untouched.

The act of felling is indecently speedy. The dangly bit on the end of the harvesting machine's arm opens a set of jaws which grab the base of the tree. An in-built chainsaw slices through the trunk and the tree falls from vertical to horizontal, still gripped at its base by the jaws. The machine's sensors then measure the diameter and height of the tree, and its on-board tech works out the optimal way of cutting up that particular tree that minimises waste. With this information, in a blur of spinning wheels, whirring blades and flying woodchips, the machine feeds the tree laterally through its clenched jaws, stripping off all small branches and stopping exactly

* I've added some suggested links on the book's website to tree harvesting machines in action. If you haven't seen one of these before, prepare to be impressed.

at the points where the chainsaw can slice the stripped trunk to the right lengths for different uses. The chopped trunks and spindly top branches are then neatly stacked, ready for collection.

Decades of growth, gone in sixty seconds.

Once these lumberjack Transformers have completed their brutal mission, the separate parts of the trees are shipped off on trucks for processing at a sawmill, leaving a landscape that resembles the aftermath of a sustained artillery bombardment. Almost every part of a tree will be used for some purpose. Even the roots and teeny branches are intentionally left behind to rot down and enrich the soil for the next generation of trees to be hacked down by the lumberjack's children or grandchildren.

Over at the sawmill, the tree trunks are de-barked and sliced into beams, planks and boards ready to be shipped off to become roofing frames for new houses, large items of furniture or freight pallets. Bark and other leftover parts are used as chippings for playgrounds and mulch for gardens. Even the sawdust and wood shavings are sold as bedding for animals or pressed into pellets for fuel. The larger branches and tops of trunks are trimmed to become fenceposts or chipped – with another terrifying blade-spinning machine – then compressed to make flooring, roofing panels and, of course, our much-loved flat-pack furniture.

Some of these woodchips are sent off to another location – a pulp mill – to take the next step in making sheets of toilet paper just like the one you hopefully still have nearby. Here the woodchips are subjected to torture by heat and chemicals to separate out the three main constituent parts of wood – lignin, hemicelluloses and the core ingredient of that sheet of toilet paper, cellulose fibres.[6] Once the cellulose fibres are separated out, they are sent to a 'pulp former', a machine which disentangles and aligns the fibres, extracts the water and spits out stiff-ish poster-sized sheets of two-millimetre-thick dried pulp. These are bundled into bales about the size of suitcases to be shipped off to a rolling mill, typically many hundreds of miles away, often in a different country.

From pulp to paper

At the rolling mill, those suitcase-sized bales of dried pulp are cut open and chucked into a machine a bit like a giant washing machine that thrashes these sheets up with water.* The resulting watery pulp is then sprayed in separate layers onto a five-metre-wide speeding wire mesh, which then carries the sprayed pulp at over sixty kilometres per hour over multiple drying stages. At the end of these stages, it is wound onto a giant steel roll from which soft, warm paper is scraped off with a giant razor blade and onto a spool.

Speed and scale? One kilometre of five-metre-wide toilet paper produced every minute, rolled onto one giant 'mother roll'.† Ply from multiple 'mother rolls' are embossed with a pattern and glued together, then chopped to the width of the paper you had in your hand earlier. Fourteen thousand rolls of toilet paper are produced every hour.

The last stage is to wrap the rolls in their outer packaging, bundle them onto pallets (possibly made from wood harvested from that forest in Scotland) and load the pallets onto the backs of trucks – each carrying fifty thousand rolls – for the next part of their long journey to your smallest room.

Distributing and retailing

Wherever you buy your toilet paper, it will have spent some time in a vast, externally uninspiring bit of architecture called a

* Yes, water is added back after spending all that time and energy at the pulp mill removing the water. Why? Removing the water reduces the weight of the pulp and hence lowers shipping costs.

† Some machines actually run at twice that speed.

distribution centre or fulfilment centre. These giant, windowless sheds tend to be clustered together at 'strategic locations' – that is, places from where it is easy to get to lots of other places. You'll have seen these massive, crinkly-façaded warehouses from the car or train grouped together in business or industrial parks. These are places where things are delivered in bulk and then sorted, stored and sent onwards with air-traffic-control-like precision in smaller bundles to other warehouses or directly to factories, shops or homes.

Since supermarkets became a common feature of our economies in the 1950s, most of us have grown accustomed to having ready access to all our staple goods in one place. Supermarkets cannot function – they cannot sell us the things we need – without the seamless integration of the activities of the hundreds of firms making and shipping all the products we expect to find on the shelves when we walk through the door. As such, supermarkets represent the very visible end-point of more supply chains than perhaps anywhere else accessible to the public.

'What we are doing', says my friend Bill, the manager of a large supermarket for one of the UK's largest chains, 'is kind of a magic show. Customers walk in and the very thing they want is just there. They have no idea how it got there, and they take for granted that it will be there. It is amazing how well this all works.'

'Rubbish!' shouts his partner preparing dinner in the background. 'The other day, I was in your shop and tried to find . . .' There followed a rant about things not being available, items being moved to different aisles and staff not being around to help find things.

Bill sighs. He is well used to comments like this, and not just from his partner. He will have been at the store at 7 a.m., and the first thing he will have done is check that the 'night fill' of the shelves has been completed properly, ready for the first customers to walk through the door. At key points during the day, he will come down from his office to the shop floor to work with his team to do the 'facing' – bringing products to the fronts of shelves to make them easily visible and accessible to customers. He knows there are certain key products that, if not in plentiful and readily visible supply throughout the day, will cause customers to get very grumpy and next time go to a competitor instead.

One of those key products is toilet roll.

Despite the protestations of Bill's partner, the fact that there are usually ample supplies of the required rolls on the shelves of every supermarket really is a kind of magic, especially when you start to think about the variety of products for sale.

It's worth a quick look at the numbers. The specific package that contained the roll of toilet paper from which you tore a sheet earlier represents what supermarkets call a single 'stock keeping unit' (SKU). All SKUs have a code number that allows supermarket managers to keep track of everything arriving and leaving the store. For example, a four-roll pack of Sainsbury's Super Soft Cushioned or a twelve-roll pack of Tesco Luxury Soft are each one SKU. The larger or smaller packs of the same products are different SKUs.

And here's what I think is a rather impressive number: thirty thousand.

That's the number of different SKUs on the shelves of an average UK supermarket.[7] That's not thirty thousand individual items, such as loaves of bread or bottles of shampoo, but thirty thousand product types, each with its own barcode. The total number of individual items in a supermarket will be in the hundreds of thousands, and each one needs to be made, delivered and then put on the shelves by Bill and his colleagues. Multiply that up across the more than six thousand regular supermarkets across the UK,* and the scale of the magic trick that Bill and his colleagues at the shop

* That's not including the 43,000 smaller 'convenience' stores nor the 7,000 forecourt stores (Institute of Grocery Distribution, 2019).

level – and all those working in retail logistics – pull off every day becomes even more impressive. Millions of individual items that have each been made and moved to arrive at the right store at the right time and put on the right shelf for you to reach forward and take when and where you need it.

As you carry your packaged rolls of toilet paper through the checkout and transport this product to its point of use, you are acting as the last link in the supply chain for the remarkable manufacturing system that connects tree to toilet.

Doesn't this all seem like a lot of effort?

The single sheet of toilet paper I teased apart earlier sits on my desk. I look at it next to the arrow-filled map I have sketched of the ludicrously complicated journey this superficially simple product had taken.

To get those rolls into those shops and warehouses, millions of trees are planted, thinned, felled, chipped and fed into a system that consumes lakes' worth of water and power stations' worth of energy to extract one ingredient. The whole system to make this product requires the brains and brawn of thousands of workers, millions of dollars of investment and the movement of materials and partly finished goods over thousands of miles. Its production requires vast machines to be designed, delivered, installed, run and maintained. Fleets of ships, trucks, trains and vans have to be dispatched hither and yon to make sure all the right things are in the right place at the right time. So much effort and consumption of resources to bring one mundane product into our lives to address one of our most basic needs.

Given the extraordinary challenge of synchronising so many interconnected activities, it is hardly surprising that things are likely to wobble a bit when confronted with a crisis of the scale of a global pandemic.

What was 'the great toilet roll grab' all about?

There were many temporary shortages of consumer staple products during the COVID-19 pandemic, but toilet paper made for easy headlines as it seemed to be something so basic. The panic buying was largely driven by the perceived threat of the pandemic and, research shows, individuals' personality traits.[8] However, you could claim that all consumers were being quite sensible. Households would need more toilet paper if everyone was going to be forced to stay at home, and it is something you really do not want to be without. As a result, just as COVID-19 lockdown restrictions were announced in 2020, toilet paper sales surged by over 700 per cent compared to the previous year.[9] Since demand for toilet paper is usually very stable, shops only keep a stock for two to three weeks' worth of sales; and not even the most flexible manufacturing system could cope with an instant sevenfold increase in demand.

In addition, manufacturers struggled with the peculiar characteristics of this surge in demand. The toilet paper industry has two distinct markets: commercial customers (e.g. office buildings, public facilities) and household consumers. While consumer toilet paper comes in smaller rolls, is commonly made from virgin fibre and is packaged in branded packs, commercial toilet paper is thicker, uses more recycled fibre and is packaged in large bundles. These two types of product are often produced in separate factories. The COVID-19 lockdowns led to a shift in demand towards consumer toilet paper as people spent more time at home and less time – or no time – out in public or at the office. In theory, mills making commercial toilet paper could have tried to redirect some of their supply to the consumer market, but shifting your product to target a different set of consumers is not like flipping a switch. It is complicated, expensive and slow, and manufacturers would have faced the same problem in reverse once people were released from lockdown and more normal buying patterns resumed.

So, 'the great toilet roll grab' was the result of perfectly rational stockpiling, amplified in some cases by the specific personality traits of customers, coupled with factories struggling to rapidly and temporarily reconfigure the making and delivery of appropriate products for different market segments.

Looking at the manufacturing challenges of just one type of product in the midst of a global crisis provides a helpful stepping stone to the discussion of a broader and very basic issue for manufacturing firms: how do you ensure that what factories produce and what customers want are always in balance?

Pulling off the toughest of balancing acts

Supermarkets provide a useful illustration of why balancing what we want against what can be provided is so difficult. When things are stable, when there is little variation, this balancing act is just about manageable. Those working in the planning departments of big supermarket chains will have terabytes of data that link those tens of thousands of SKUs with the past buying behaviour of their millions of customers, allowing them to predict with a pretty high level of confidence what customers will want and when they will want it. The purchasing teams at supermarkets can work with the vast number of external suppliers and their own logistics teams to ensure that all shops have just enough stock to satisfy their customers. Under these conditions, supply and demand can be pretty tightly matched.

But there will almost always be a bit of variation.

That could still be OK when it is predictable variation. For example, demand and supply for certain products will fluctuate with the seasons. Or there could be a major planned event like a seasonal holiday, a sporting event or a royal wedding; supermarkets know that customer preferences will change for that event and that suppliers will need to ramp up or ramp down their output before, during and after. That makes things a bit more complicated, but still broadly manageable.

But then the error bars on those predictions widen. Sure, supermarket managers know summer is coming, but will it be a cool or hot summer? Will it be wet or dry? These issues will affect not only how customers behave but also what suppliers can provide. Again, though, the supermarkets use some amazingly sophisticated tools to draw insights from the vast amounts of data they can now access on everything from weather predictions to sentiments picked up from social media posts.

However, there's another layer to the unpredictability against which it is much harder to plan and respond rapidly: how do you factor in the impact of a war, or an unexpected election result, or a global financial crisis?

Those involved in risk management will typically look at two things related to such unknowns: likelihood and impact. Very few businesses can afford to invest too much preparing for events that have an incredibly low likelihood of happening, even if the associated impact would be high. Just think about how much it would cost for any firm to be 100 per cent prepared for a future global pandemic.

Realistically, it's just not worth it.

Uncertainty has consequences across the three broad areas of activity within the manufacturing world: making, moving and consuming. To illustrate this point, I want to stay in the supermarket but take you away from the fragrant scent of the personal hygiene aisle and over to the coolness of the fresh food aisle.

In manufacturing, even the simple is complicated

I didn't think that lettuce could be classed as a manufactured product until I spoke to my colleague Mukesh. He specialises in trying to understand why some manufacturing systems don't cope well with various forms of disruption, and he uses this knowledge to help companies and governments do things to increase the resilience of such systems.

When I asked him to share an example of his work, somewhat to my surprise, he told me about lettuce. The picture he painted was way more involved than I had anticipated. The lettuce supply system has five steps between 'field and fork': farms, cooling facilities, distribution centres, customer outlets and customers. Within each one of those steps, there are multiple stages of activity – at the farm, for example, there is the land preparation, planting, cultivating and harvesting (which itself includes the multiple steps of cutting, trimming and packaging). Farms have three types of customer: the wholesalers who buy from farmers and sell on to the shops, the shops who buy directly from the farmers, and the processors who buy in order to do some additional 'value add' (chopping, mixing into salads, packaging into bags) before selling on to shops.

So, what sort of disruptions might this lettuce supply network face?

First, there are the obvious ones like huge variations in weather patterns. The consequences of this are clear. Crops are chosen for their ability to thrive in certain climate conditions. If those change, yields drop. But that's not just a supply problem. As we know, demand may change depending on the weather. Lovely summer? More barbecues and picnics push up demand for salads. However, very hot weather is bad for growing lettuce, so in a heatwave, supermarkets may have to import more from cooler locations.

Systems like this are dependent on much more than just weather. For example, following Brexit, the supply of manual workers coming to the UK to harvest and transport lettuces started to dry up. Meanwhile, with the war in Ukraine and associated sanctions imposed on Russia, supplies of AdBlue – essential for running diesel engines cleanly – ran short. Why? Because energy price hikes had made it uneconomical to produce.[10] Without AdBlue, modern diesel engines won't run. No diesel-powered trucks, no way of getting lettuces from farm to distributors and retailers.

The example of lettuce production and distribution shows us some of the challenges of balancing supply and demand in the real world where things are in a constant state of change, uncertainty and unpredictability.

But what if the product is a bit more complicated than a bagged lettuce? What about a car with its fifteen to thirty thousand components?* Or an airliner with roughly six million?[II] If it is a bit tricky to ensure the consistent supply of lettuce and loo roll, just imagine the headache of trying to ensure the resilience of the supply networks for all the components needed to make cars and planes. Yet, somehow, manufacturers manage to synchronise the making and moving of those thousands or millions of parts needed to get cars on the road and planes in the sky.

How is that complexity managed?

In order to have any chance of managing things, you first have to be able to know what is going on across the whole web of suppliers. And, as another colleague from the Institute for Manufacturing was about to explain to me, when you are dealing with supply networks for cars and planes, that is very far from straightforward.

To manage a complex system, you first have to visualise it

It was a picture-perfect early autumn morning as I steered my bike over the satisfyingly crunchy leaves that had blown onto the cycle path. Waiting for me at our institute on the campus to the west of Cambridge was my colleague, Alexandra. She had promised not to laugh when I asked her some really basic questions about automotive and aerospace supply networks.

* Depending on whether it is an electric vehicle or an internal-combustion-engined one.

Alexandra's research explores how we can visualise and understand what is happening in the most complex of supply networks. She has no shortage of funding, as many big manufacturing firms need to understand their own networks better. They will, of course, know who their immediate suppliers are, but as you descend deeper into these networks, you'll have less detailed information on your suppliers' suppliers . . . and your suppliers' suppliers' suppliers . . . and, well, you get the picture. And if you don't know who is buried down at the base of those networks – and what problems they might face – how can you assess the risk to your business should one of these suppliers fail to deliver?

Sitting with Alexandra in our institute's common room, I try to focus on an image she is showing me on her laptop. The picture is a kind of mad multicoloured spider-web, combined with one of those Magic Eye things. As I polish my glasses, she is patiently explaining what it represents: a map of all the organisations involved in making one product (a car in this case). Each dot is one firm, each line a trade connection between one firm and another.

'Complex network' is a fair description of what is going on, and complex networks are, well, complex. But that word has a particular meaning here. Complex means that the dots are *interdependent* – something that affects one will affect others. And when you have lots of interdependent dots – or nodes – in a network, it is really hard to predict what will happen to the whole system when something happens to any one of the firms represented by one node. When things do change, the impact can spread very fast.

That is what has been happening increasingly frequently in the manufacturing world.

A node could represent one material supplier for a minor component in subsystems that form part of a slightly more important system that is part of something essential for the operation of a car, plane, phone, medical device or satellite. If something goes wrong at the first node – the material supplier – you want to know what the likely impact is going to be on the rest of the system. But things tend not to happen neatly, one at a time. What if several nodes across the network each face a disruption – a flood at one, a strike at another, a CEO getting sacked at another? What will be the impact on the overall system of those nodes facing simultaneous crises? These are issues even the most experienced manufacturing brains will struggle to cope with. And that is why more and more manufacturers are looking to AI to help them out – and why Alexandra's lab is called SCAIL – the Supply Chain AI Lab.

The unintended consequences of being too efficient

There is another fundamental problem that makes these complex manufacturing systems more vulnerable to shocks, be they producing salads or toilet paper, cars or planes. We have inadvertently made them alarmingly fragile by trying to do a really good thing: reduce waste.

The supply networks being researched by Mukesh, Alexandra and their teams have been refined and optimised over many years to be as efficient as possible. Often this has been through the application of what is called 'lean manufacturing'. One of the key principles of this approach is the reduction or, ideally, elimination of all forms of waste. In the world of lean manufacturing, 'waste' is any activity that consumes resources but doesn't add value to the end customer. Under that definition, waste includes things like keeping some extra parts in the storeroom, or producing a few more things than you actually need 'just in case'.[12]

Reducing waste means that costs can be minimised and hence prices kept low for us as customers. But if this is taken too far, you end up in a bad place. There are no back-up supplies in the back of the store to use when there's a glitch in production; you don't have multiple suppliers in case one has a problem, and instead you focus on operating things efficiently with as little inventory and few suppliers as possible. The result? Low-cost, high-quality products available very rapidly (which we all like). However, this relentless focus on efficiency can leave manufacturers with a much-reduced ability to respond when things go wrong – which, at some point, they inevitably will.

Every node and every connection on one of Alexandra's supply network maps represents a risk. In the quest for low costs and immediacy, we have built systems that have no Plan B. But you can kind of see why manufacturers have done this. Who really believed that a 200,000-tonne ship with a capacity of twenty thousand TEU* containers would get itself wedged in the Suez Canal, resulting in over a tenth of all global trade between Europe and Asia being stopped? Or that a real, bloody, protracted land war could break out on the eastern edge of Europe in 2022, driving energy and crop prices through the roof? And how likely did a global pandemic and national lockdowns seem before 2020?

Such events have made visceral something that is normally obscure and remote. The global systems for making and delivering the things we need to function on a daily basis started to creak, and we didn't like the sound. Then, when they snapped, the consequences of our blinkered understanding became all too clear.

Each recent crisis has amplified the necessity – and widespread desire – for us to take a good hard look at how products get made, delivered and consumed. And, as we shall see, this isn't just about ensuring that we have manufacturing systems better able to cope with occasional disruptions. It is also about ensuring they do not inadvertently destroy our ability to inhabit this planet.

* TEU = Twenty-foot Equivalent Unit, the standard classification of the size of a container = twenty foot long, eight foot high and wide. The long ones that take up the whole of the back of a truck are typically two TEU size.

This brings us near the close of this miniature version of the rest of the book. We zoomed in to explore one small corner of the manufacturing world. By taking you on a sprint through the whole make-move-consume journey for one of the most mundane of products, I hope you can see that even the manufacturing of something as superficially simple as a sheet of toilet paper requires the co-ordination of a remarkably complicated set of activities.

We then zoomed out to look at some of the characteristics of the wider system. This system is simply extraordinary in the variety and number of products it can now make and deliver to attempt to slake our seemingly unquenchable thirst for more and better things. But we've also seen how this system has developed two emergent properties we are less happy with: it is mind-bogglingly complex and worryingly fragile. If we are to avoid suffering the worst consequences of these two unwelcome features, we, as consumers, need to be aware of our roles within this system and the consequences of the demands we make of it.

And that is exactly what the following chapters are going to help you do.

However, to understand what is going on, some simple tools, concepts and terms will be helpful.

To provide you with the first set of these tools and to add the first bit of detail to our picture of the manufacturing world, I can't think of a better place to start than with some chocolate.

2: Make

What actually happens inside a factory?

The gold envelope is torn open and the card removed. A pause. I pretend not to care what the next words uttered by Bridget, the chair of the judging panel, are going to be.

The certificate still sits proudly in my office. Winning a prize at our institute's annual charity baking competition (award sub-category: tray-bake) has been a definite career highlight. The creation that won me this prestigious award was a tray of chocolate brownies, baked, dusted and packed in a rush the evening before, then cycled to work in a bag dangling from the handlebars. This trivial example of production will help answer the very simple question: what happens inside factories – the part of the manufacturing world where things are actually made?*

To answer that question fully, I will need to take you on a journey to visit a range of different factories. However, without an understanding of some of the techniques of production, just looking inside some factories might not reveal that much. It'd be a bit like watching an action movie in a language you don't speak, without subtitles. You'll get a sense of the drama and excitement but won't really understand why those people are shouting at each

* I'm going to be using the very broad definition of 'factory' in this book, i.e. 'a building or buildings where people use machines to produce goods' (https://dictionary.cambridge.org/dictionary/english/factory), rather than the narrower definition of a place where *lots* of things are made.

other, the significance of the clown or whose head they've just found in the bin.

Thankfully, I have some good news. Though I said at the start of this book that the world of manufacturing has its own at times daunting language, some of it is really simple and clear. To be able to interpret what is going on inside a factory, I reckon you just need seven words. All factories do the same thing: they take some *inputs* and do some kind of *processing* (*people* follow a *method* using *machines* and *materials*) to convert them into more valuable *outputs*. That's the whole production part of the manufacturing world summed up in seven words.

And it doesn't matter what is being made, those basic activities are the same. You have likely seen video clips of factories showing rows of workers making thousands of phones in vast city-like factory compounds, or millions of components being assembled in cavernous hangars into an airliner, and you might have thought, 'That looks quite complicated.' And it is. But the fundamental building blocks of methods, machines, materials and people brought together to run through the cycle of input > process > output are the same. So, if we want to understand the basics of what happens in a factory, it doesn't really matter what product we look at – and that includes a tray of chocolate brownies.

A small domestic food factory

This factory is quite small: three metres by four. It contains two worktops, an oven, a stove, a fridge, some cupboards, a dishwasher and a frankly ludicrous number of tools of various sizes spanning the frequently to the never used. Some are simple manual tools, some are powered – and of those, some require my direct involvement, and some are partly or completely automated. There is one worker (semi-skilled), combining the roles of supervisor and shop-floor operator.

I pull out a recipe book from the shelf next to the fridge and zigzag between the cupboards, fridge and worktops, assembling everything required. I now have in front of me all the raw materials and tools I think I need.

Production can begin.

I heat some water in a small pan then place an alarmingly large amount of butter and dark chocolate in a Pyrex bowl, which I then plonk on top of the pan simmering on the stove. While those two ingredients slowly begin to melt, I weigh an equally concerning amount of caster sugar into a larger glass bowl and break in two eggs. I watch as the sugar and egg transform from granules and goo into a smooth creamy liquid, listening to the tone of the electric whisk changing as the mixture becomes less viscous. I return to the stove and poke at the chunks of chocolate and block of butter in an ineffectual attempt to speed up the melting process. I make a cup of coffee to fill the time and drink it while losing a staring competition with a squirrel sitting on the fence outside.

The warm smell of melted chocolate snaps me back to the task in hand. The mixture in the heated bowl is now a glossy dark brown liquid, ready to be added to the eggs and sugar. I grab a cloth, awkwardly pick up the bowl by its rim and walk slowly and steadily from stove to worktop. I add the melted chocolate to the big bowl and watch as the two swirling loops of dark chocolate and creamy, eggy sugar are whisked into one caramel-coloured whole. I almost drop the slippery hot bowl gripped in my non-dominant hand. Blending done, the recipe next tells me to sift in flour, cocoa powder and a magical additive called xantham gum.*

Then, crisis.

Opening the tin of cocoa reveals only dregs. Annoyed at my own lack of preparation, I grab my keys and coat and stomp out of the house to visit my local mini supermarket. Breathlessly returning

* Apparently useful for both gluten-free cooking and putting up wallpaper.

a few minutes later with the missing ingredient, I pick up where I left off. I sift in the cocoa and gently fold everything together, adding orange zest to give a false hint of healthiness.

The now slightly gelatinous mix is poured into the lined baking tray, which I then shimmy to help the mix settle more evenly. The tray is placed in the oven. I spend the baking time on Duolingo, surprised as ever at the phrases the developers think would be useful to slip into future conversations.*

My phone timer sounds, and I bend to peer through the slightly grubby oven door. Perfect. The mix has risen, creating a beautiful stretched shiny surface. Opening the oven, I'm hit by a rush of steam that fogs my glasses, leaving me blinded and gormless. As sight returns, grabbing the nearest cloth to protect my hand as I fumble in the oven for the baking tray, I turn and almost throw the hot tray onto a cooling rack as the heat through the thin cloth becomes unbearable. A soft creaking and popping sound accompanies the cracking of the surface of the mixture while I cool my fingers under the flowing tap.

I push on to complete the final two tasks: I dust the hot brownies, my jumper and everything nearby with icing sugar, before slicing them into equal-ish pieces for serving. My hand reaches for a perfectly formed corner piece, but then I remember that these aren't for me; they are for tomorrow's baking competition.

With a deep sigh, I reach instead for a plastic container to seal the goods away from temptation.

For any manufacturing engineers reading this book, please accept my apologies. That flippant example was probably quite painful to

* A recent Japanese favourite being 'あなたの顔はこのジャガイモに似ている' – 'Your face resembles this potato.'

read. But I want to use my obvious amateurism to reveal to everyone else how things should be – and are – done in real factories.

The value of mapping things out

One of the first things we get students to do when they join the institute is spend time in factories undertaking something called '*value stream mapping*'. It's a simple but hugely useful exercise: we ask them just to watch what is happening, ask questions and dig out data, after which they draw a picture[1] – a map – of the flow of people, things and information needed to complete a process. This is a great way of understanding what is going on, but it also provides an opportunity to spot where flows and tasks aren't quite as smooth or efficient as they could be. For example, if I were to ask the students to watch me baking, they could quickly observe the time-consuming toing and froing (how many times did I walk between the cupboards, fridge and worktop?), highlight areas with potential risks (why am I turning and walking across the kitchen with a slippery bowl of melted chocolate?) or point out preparation issues (why hadn't I checked the cocoa supply before starting?).

These maps often raise issues relating to how parts of the factory are laid out and where various things are positioned. For all sorts of reasons – lack of space, access to power, too difficult to move or 'it's just always been there'* – machines and tools might not always be in the best location. Some equipment might be movable (I could move the bowls, whisk and spoons a bit closer to the stove), but some might not (it'd be tricky to move a built-in oven to bring it closer to the worktop). The students quickly get to see that there are often things that could be done better,

* There's even a special name for bits of kit like this that are just too difficult or expensive to move; things such as furnaces, ovens and rolling mills. Due to their implied permanence, they are known as 'monuments' and they may end up dictating the layout of the rest of the factory.

but that there may be very good reasons for why things are as they are.

In order to understand why things are happening in a particular way, we need to look at each of these recipe steps in a bit more detail. What we need to do is to explore how a manufacturing recipe – or what engineers call a *process design* – is developed. To do that, let's briefly step back into the kitchen.

The starting point for any process design is answering two simple questions. What is the product? How many are going to be made? We'll come back to that second question later, but for now I want to focus on how the nature of the product dictates the design of the process for making it.

A manufacturing engineer given the task of writing a recipe (process design) for baking brownies would begin with the final product and then define all the different 'states' it will have had to go through. For this example, the states are: raw materials prepared for mixing; tray of brownie mix in liquid form; single slab of baked brownie mix at room temperature; individual sliced and dusted brownies in a plastic pot.

Next, the engineer must work out what methods* are needed to convert the raw materials – or, as things progress in semi-finished form, the 'work in progress' – to the next required state. There could be multiple options for each of these transitions, or 'unit processes' (I could leave the butter and chocolate sitting in the sun rather than over a pan of hot water), and each option will have pros and cons (leaving things to melt in the sun will be quite slow, especially in winter in England). Whichever option is chosen for each unit process, it should not change the final output.

* I could fill up the rest of this book with descriptions of the different methods that manufacturers use for transforming materials from one state to another. Instead, I've placed some links on the book's website (www.timminshall.com) for those who'd like to know more.

All of these unit processes are then connected together. This is quite simple for something as basic as baking brownies, but when you are working on something slightly more complicated – like a laptop or an aircraft – it requires a bit more thought.

Then, there are choices to be made about what tools and machines are going to be needed. This will be linked to another key requirement: what roles will people be playing in this process? For each unit process, do you need highly skilled people working with simple tools, or are you going to use more automated equipment that doesn't need such high skills to operate?

Next, you want to make sure that everything is optimised to help achieve a smoothly flowing set of activities. To achieve flow, you need to make sure that every machine and person has everything they need to carry out their tasks. And to ensure that, you want to position each piece of equipment in its optimal location, relative to other relevant bits of kit, and to ensure that all the materials or components are delivered to that place when needed and in the right form. You'll also want to make sure, once everything is up and running, that you can monitor what is going on to spot anything that is not going quite right and needs attention.

Finally, you would normally want to test the whole process to make sure everything works as it should – delivering outputs at the desired quality – before you make a real product that you will be selling to your customers.

This simple description of baking has shown some generic features of the way things are made in any factory: methods, machines, materials and people are co-ordinated to take basic inputs and convert them into more valuable outputs.

But my domestic factory has one other important feature: it is designed for very low-volume production.* In this facility, I produced just one tray of brownies, and, frankly, I'd struggle to make many more. What happens when you want to produce more of a particular product – *much* more – every day?

Baking in batches

It is a cliché, but I'll say it anyway: Fitzbillies bakery has become a Cambridge institution. Despite setbacks and restarts (including due to a fire, a bankruptcy and multiple changes in ownership), they have been making and selling bread and cakes since they first put the 'Open' sign on the door of their Belgian Art Nouveau-style shop on 4 October 1920.[2] The current owners are Alison Wright and Tim Hayward, who took the business out of bankruptcy in 2011. They've done a remarkable job and the business is now thriving.

At its heart is a well-run factory operation, and Alison has invited me to see it in action as it produces hundreds of its most famous product: the Chelsea bun.

My alarm wakes me 3 a.m. Half an hour later, I'm cycling through the dark empty streets, dodging the infamous Cambridge potholes to an industrial estate behind the railway station.

Walking inside the bakery, with its warm, sweet, yeasty aroma, I immediately feel like I am in the way. There is so much crammed into this relatively small space. Every bit of floor-space seems to be occupied by people, equipment or materials. Some pieces of equipment – like the mixing machines and ovens – are recognisable as jumbo-sized versions of things in my kitchen, and the ingredients are the same but in much larger bags and pots.

* Though it can cope with very high *variety*. Depending on the skills of the workforce, it could be used to make any of the recipes detailed in the books on the shelf, or anything spotted on Instagram.

Three metal tables dominate the main space. On one, large loaf-like shapes of dough are lined up while they are 'proving'. On another, bakers Sam and Toni are rolling out proved dough into sheets around 150 centimetres long and 50 centimetres wide. Onto this, the special sweet brown paste is spread, followed by a sprinkling of currants. Each sheet is then rolled into a long sausage and droopily carried by hand over to Jean who is standing at the third table positioned next to the ovens.

There, with impressive speed and accuracy, Jean chops each long roll into exactly forty individual swirl-profile buns with almost zero waste at each end. These are then packed into lined baking trays in eight rows of five, and each tray is slid into the stacked rails of a tall trolley which is rolled into one of the floor-standing ovens. Timers are set, and Jean returns to his next task. After baking, the trolleys are wheeled back to the same table where the dough was rolled. Once the trays are offloaded from the trolleys, Toni dips a large brush into a bucket of syrup and paints it over the contents of each tray of cooling buns. At the door is Steve, patiently waiting to take the first of the glistening, wonderfully scented trays out to his van for local delivery; others are kept back to be boxed up and shipped to customers around the world.

The way in which these Chelsea buns – as well as nearly all of Fitzbillies' other products – are made is called *batch production*. The bakery is set up to produce batches of a range of standardised products. While I tried not to get in the way, Sam, Toni, Jean and their colleagues were seamlessly switching between the production of batches of different products – before the sun had risen, they had also made croissants, focaccia and hot cross buns.

But upstairs, there is another, smaller part of the business that illustrates a different type of factory operation known as a 'job shop'. Here, customised items like wedding and birthday cakes can be made to order, and the way this operates is like a very professional version of any domestic kitchen, with standardised equipment being used to make single, one-off products.

Baking at mass scale

Fitzbillies make quite a lot of Chelsea buns – around eight hundred on a typical day. But compared to the output of bakers like Premier Foods, owner of the popular 'Mr Kipling' brand, they hardly register on the same scale. The Premier Foods factory in Stoke-on-Trent produces over a quarter of a million cherry Bakewells *every day*.

Does making lots more of something change how things are done in the factory? If Alison wanted her team at Fitzbillies to start baking thousands rather than just hundreds per day, what would need to change? She could just recruit lots more bakers, add some more zeros to the orders from her suppliers and power up lots more ovens. But that would be hopelessly inefficient. The costs would be exorbitant, and it would be very hard to maintain consistent quality. The way in which things are organised for producing things in very high volumes – *mass production* – requires a different approach, one that is built upon a single, fundamental requirement: standardisation. If you want to make lots, make them all exactly the same. And if you are doing exactly the same thing over and over again, machines that precisely execute those repetitive steps are usually what are needed.

There's something very particular about big food factories that produce very high volumes of standardised products,[3] and they share some common features regardless of what they are making: they are always reassuringly clean with an at times confusingly blended scent of cleaning products and foodstuff; there is always lots of bare shiny metal and blue plastic; and they are always populated by people wearing really-hard-to-look-cool-in protective workwear.

In these factories, automation is very evident. Each stage of the process design, the recipe, has been refined and optimised before being embedded into a set of operations for hypnotically repetitive specialised machines to complete. These machines are connected by networks of conveyors and chutes – plus the occasional human hand – to ensure no pauses in the flow of production. This orchestra of rolling, cutting, baking, squirting, picking and placing machines minimises waste while ensuring consistent quality.

People simply feed and maintain the machines.*

Continuous making for baking

There is another type of factory that can be explained with a baking-related example, one that produces a single product but in huge quantities.

Dotted around East Anglia, easily spotted from afar by their plumes of steam and giant storage silos, are a small number of large factories whose main output is a staple for all kitchens and bakeries: sugar. In the UK, much of the sugar we consume comes from locally grown sugar beet. The process of converting thousands of tonnes of muddy tuber into thousands of tonnes of pristine sugar requires high levels of energy for heating things up, lots of machinery for moving and chopping materials, some clever chemistry and, in a modern facility, surprisingly few people. A sugar processing plant is an example of a *continuous flow* process – it is not producing discrete items but rather producing material in bulk.† Each one of the silos in a sugar factory I visited recently

* But there always seem to be some things that machines struggle to do better than humans. Before they head off to packing, at the Premier Foods factory the final addition of the half glacé cherry on top of each of the 250,000 cherry Bakewell tarts was still – the last time I checked – done by hand.

† Well, I suppose technically each grain of sugar could be thought of as an individual product, but you get my point.

contained twelve thousand tonnes of sugar. And there were seven silos. That's quite a lot of bulk.[4]

The continuous flow line does just what it says: raw material is converted chemically, thermally and mechanically, without pause or interruption, into the desired processed material. And, in this case, that processed material is then poured into sacks or bags, which are shipped off to commercial bakeries like Fitzbillies and Premier Foods or to supermarkets for domestic use.

It's all about trade-offs

Manufacturing – like so many things in life – is about making trade-offs. If I want X, I can't have Y; if I can have Y, it's going to cost me a fortune, or I'll have to wait a long time, or it'll be a bit rubbish. One of the main trade-offs is between how many things you want to make (volume) and how many different things you want make (variety). The four examples of my own kitchen, Fitzbillies bakery, the Premier Foods factory and the British Sugar processing plant illustrate how the trade-off between volume and variety affects the design of a factory.

If I wanted to make a different cake in my kitchen, there's not really a problem. My kitchen is designed and equipped to allow pretty much any basic recipe to be followed. However, a domestic kitchen is not designed to make things in high quantities – I'd soon run out of space and things would get clogged up and confused. On the other hand, if I were to ask Premier Foods to make me a personalised birthday cake, they'd struggle. The process that has been optimised in the factory is specifically for making the same things over and over again at high speed and low cost. Sitting in between those two is Fitzbillies. They combine the ability to make the same thing over and over again (e.g. their Chelsea buns) with the capacity to produce individual bespoke products (e.g. wedding and birthday cakes). Then, there is the world of sugar processing. The sheer scale of system required to do that means that you need to have a single, very large, dedicated factory for making sugar. And that factory can't really do anything else.*

Apparently, there's more to life than baking

This journey through the world of baked goods production has provided a flavour of how things happen in one key part of the manufacturing world – the factory. You now know how the nature of the thing you are making – and how many you want to make – is linked to the design of the factory where it is going to be made, and how factories are designed to some pretty standard templates.

To reassure you that the principles illustrated in my kitchen, Fitzbillies bakery, Premier Foods and British Sugar are pretty standard to nearly all factories, I want to take you on a visit to a few places where non-baked things are made.

Job shop – a blacksmith

The job shop – with the domestic kitchen being one example – is the oldest and simplest form of 'factory'. Someone buys equipment and develops appropriate skills and combines them to make anything that matches their capabilities and the customer's budget. A blacksmith used to be one of the most common of these. Need shoes for your horse, or a metal latch for your door, or a thing to poke the fire? Just wander down to the local blacksmith and discuss what you want. A blacksmith has a set of what are called general purpose tools – such as hammers, an anvil, a forge – and applies these to make whatever has been asked for.

* Although, as we shall see in Chapter 8, there are some interesting things now happening in the world of sugar processing.

In the village in Norfolk where I grew up, there was a blacksmith's firm called Thomas B. Bonnett's.* While I didn't have a horse that needed shoes (or even one that didn't), sixteen-year-old me did have a moped with a broken suspension bracket. Solution? Sketch the thing I needed on a scrap of paper and have a chat with the owner, John. I went back the next day and he'd made exactly what I needed. While you might be thinking that this seems a bit old-world craft-y, this model of production can be seen in all sorts of places from high-end jewellers and bespoke carpenters to, as we saw earlier at Fitzbillies, birthday and wedding cake-makers. The job shop is still a core feature of the modern manufacturing world, and it is one that has flourished thanks to the internet: think of websites that offer individually customised T-shirts or personalised greeting cards. For units of one, the job shop remains the factory design of choice.

But as Fitzbillies bakery demonstrates, if you want to make more than a one-off, unique product, you need a factory that is suitable for producing batches of the same thing. To illustrate this, I want to take you to a production facility that makes a batch of just two eye-wateringly expensive, precision-engineered products per year.

Batch production – a racing car factory

A high-end racing car factory is quite special. For those making cars to compete at what many regard as the pinnacle of motorsport, Formula 1, they are *very* special. They are part space-age lab, part steampunk workshop, and often have a museum and giftshop attached. They may have jaw-droppingly large budgets to bring their extraordinary products to life, but these factories share a lot in common with any other batch production facility.

* And there still is, as it has passed through five generations of the family since it began. See the company history at https://www.thomasbbonnett.co.uk/about-us.

Product design is constrained by the rules and regulations of the Fédération Internationale de l'Automobile (FIA), motor-sport's governing body. Every year, the rules for what the teams are allowed to build will change, and so F1 teams must align their people, materials, equipment and methods with whatever new regulations apply. To prepare for this, they equip their factories with some of the latest and most exotic production technologies, from high-end computerised cutting machines for creating exquisitely complex metal components to autoclave technologies needed for curing the composite materials that provide maximum strength with minimum weight. Because this is batch production, and because the product varies each year, this is a very people-focused operation, and those people are amongst the most talented engineers money can buy.

As I walked through one of these facilities recently, the word 'factory' didn't seem appropriate. Yes, the metal machining and carbon fibre composite production areas don't feel that dissimilar to many other small modern factories, but the difference is that everything here seems very shiny, very new and undoubtedly very expensive. The assembly area has two U-shaped bays within which sit part-finished . . . things. It doesn't feel right to call them cars as they're wheel-less at this stage, all wings and curves shaped around the sunken cockpit and bulging engine. This part of the factory has the vibe of an operating theatre – it even has a glass wall separating visitors from those wearing sponsor logo-laden workwear, labouring to bring this batch of two machines to eardrum-bursting, blisteringly speedy life.

If that racing car factory is the automotive equivalent of Fitzbillies bakery, I want to stick with the automotive theme and show you the car-producing version of the Premier Foods factory. In fact, I want to take you to one of the newest and most advanced.

Mass production – a car factory

The Zeekr electric vehicle factory in Ningbo, eastern China is very new and very shiny – and very, *very* big. The site covers eight square kilometres and has the capacity to produce 300,000 vehicles per year.[5]

A golf buggy propels me through the factory, making me feel just a tad self-conscious, but this mode of transport provides a great way to follow the flows of materials and components as they progress along the production lines. And you get to see clearly how many of the activities have been automated, and where there is still the need – for now – for manual input. At one end of the factory, where the 'megacasting' of the main chassis components takes place, there is near 100 per cent automation of this hot and heavy process; assembly and joining of the main components of the cars is also largely done by robots. As you move further along the line, however, where there is more complexity, dexterity and variety required, the level of human input increases. Scooting around the whole site is an army of little automated guided vehicles, carrying out the myriad tasks needed to ensure the people and robots have the right tools and materials to do their job, and that everything continues to flow smoothly.

The Zeekr plant is at the top end of mass production. It has been designed specifically to make high volumes of high-end electric vehicles. And, as we shall see when we return to look at this factory in more detail, it is doing some very clever things to try to reach that manufacturing nirvana of *mass customisation*: combining high volumes with high variety.

Continuous production – the paper mill

For the non-baking-related equivalent of the continuous flow process shown at the British Sugar plant, we can return to the earlier example of a paper mill.

Paper-rolling mills are basically one huge machine wrapped in a building. The numbers associated with the top-end paper-rolling machines are staggering. The largest of these machines – just one machine – is the length of six football pitches and weighs thirty thousand tonnes.[6] At full speed, the fastest of these machines can spit out over two kilometres of ten-metre-wide paper *per minute*.[7]

When I read these numbers for the first time, I just couldn't process them. What does paper travelling at over 130 kilometres per hour look and sound like? How long does it take to walk from one end of these machines to the other?

Luckily, there was one of these located about a ninety-minute drive from our institute, so I went to see for myself.

My host at Palm Paper, Ian, walked me right through the whole, very linear process. We started at the dusty end. As this facility uses 100 per cent recycled paper as its input, it first has to sort and process used paper. Trucks offload tatty bundles of paper, which are plonked onto conveyors that pass this feedstock through a series of sorting and filtering steps. These first steps are vital to ensure no non-pulp debris – like plastic or staples – ends up entering the delicate innards of the massive machine around which this £400 million facility has been constructed.

We walk up and down metal walkways to a rumbling, roaring soundtrack deadened by earplugs, following the journey of the increasingly purified, de-inked, mashed-up pulp. We eventually emerge into a largely people-free aircraft-hangar-like space that houses the PM7 paper-rolling machine itself – though the word 'machine' doesn't feel right. The PM7 is so massive that it feels more like a piece of architecture than a piece of equipment. Stretching from floor to ceiling and along the whole length of the building, the PM7 roars like a jet engine at cruising speed. Walking along its length, I note that many of its inner workings are shielded behind metal covers, occasional windows in the inspection doors revealing the blurred spinning of giant rolls of increasingly refined paper. As we near the end, huge yellow overhead gantries lift away the giant finished spools of bright white paper.

We take a detour to enter the relative quiet of the control room. Here, banks of screens display data captured at every point, from scrappy, ink-covered paper entering at one end to finished rolls of pristine paper robotically offloaded to dispatch a few hundred metres later.

With so much variability in the quality of the input waste-paper, the PM7's vital signs need constant monitoring by a team of specialist engineers, its settings being tweaked to ensure that none of that variability of input messes up the standardisation of output. Given that this process is running at two kilometres per minute, Ian's colleagues' attention is totally focused on the numbers and colours on those screens: they really need to spot things early and act quickly. Ian gently ushers me out of the control room before I inadvertently distract them from spotting a vital blip in data that triggers a production disaster.

Like the British Sugar factory, the Palm Paper site embodies modern continuous flow manufacturing: massive scale, lots of automation, relatively few people, massive energy requirements and a relentless focus on ensuring consistency of output and reduction of waste.

So now you have the four non-baking examples of the four basic types of factory. But there is one other type of production facility that doesn't have a parallel in the world of baking. This is a facility that makes things in very low volumes (often just one) but very high variety (almost every one is unique). And it's one of the strangest types of factory. Imagine this: a factory that is assembled, makes one single product, then disappears. It only exists while the product is being made.

And Sam – project co-ordinator at the French construction company Bouygues – invited me to see such a factory from the inside.

The manufacturing project – constructing a building

After an admin-heavy week, a visit to something 'real' seemed like a good way to segue into the weekend. And things don't get much more real than a building project of this scale: the £300 million new home for the University of Cambridge Department of Physics.[8]

Running slightly late, I half-jogged across the West Cambridge campus to the solid wooden fencing that sealed off the construction site. Reaching a gap in the fencing, I peered through to see a laminated A4 printout: 'Site Entrance →'.

I walked in the direction of the arrow on the high-grip plastic honeycomb matting sitting over the mud until I reached a turnstile. A quick call and Sam was there to greet me and let me through to reveal the workings of this rather extraordinary type of 'factory'.

As this project was at the later stage of its construction, some of the artefacts of this temporary factory were already disappearing. The project team's Portakabins that had been one of the first things to appear four years earlier were being craned away – shortly followed by the cranes themselves – to form the start of another project elsewhere in the country. By this stage, the project team had been decanted from those temporary structures to become the first residents of one of the vast unfinished spaces within the building. In about a year's time this particular space would become a thrumming workshop building the exotic machines needed to test the theories that could explain the mysteries of the universe. For now, though, the cement-scented space was filled with procurement managers chewing the ends of cheap biros as they worked out which part of a form needed to be completed to confirm the safe arrival of a shipment of door handle fixing screws.

As I walked round the network of connected, half-finished buildings, the different stages of the factory dissolving away were clear. On the upper floors of one wing, some of the offices for researchers were finished, sealed and awaiting furniture; on office doors, taped-on notes threatened that 'Red Cards' would be issued to any worker dumping tools or generally doing anything in that room that might scuff the paintwork or dint the plaster. Elsewhere, temporary, loosely hung chipboard doors separated spaces where things were definitely still more factory than product: hoists and scaffolds positioned workers and equipment needed to complete wiring and plastering, while temporary workbenches allowed carpenters to construct the framework for the tiered lecture seating. Over the next few months, all these artefacts of production would be gone, leaving only the product itself: the building.

As Sam escorted me off the site, I asked her what she would be doing once this project was complete. 'Oh,' she said, 'I've already started on the next project, one down in London.'

What seemed to me like a production process with a final endpoint of completion is just – for those on the inside – part of a continuous cycle that blends factory-build to product-build to factory-gone.*

Factories are reflections of the products they make

Those examples of construction (project), village blacksmith (job shop), racing car construction (batch production), EV car assembly (mass production) and paper production (continuous flow) reinforce

* Timelapse videos of construction projects are a really helpful way of making this point clearly. I've put some examples on the book's website – including one of the projects that Sam showed me around.

the baking-focused point made earlier: what you want to make and how many you want to make will define the type of factory needed.

The manufacturing world is a giant interconnected network of millions of factories built to these different basic templates. Take that Zeekr car plant as an example: it was a project to build the factory itself, with customised fixtures and fittings produced in job shops. The robots that now populate its shop floors will have been made by suppliers producing in batches. The mass-produced cars themselves will have been assembled from the outputs of the hundreds of other factories producing and delivering batches of everything from wheels to wiring to windscreens, which themselves are produced from the outputs of continuous flow lines making all the metal, glass and plastics from which every car is made.

The design, location and operation of factories is continuously evolving, and some truly extraordinary things happening now are pointing to a surprising future for the factory. However, to understand what might be lying ahead, we first need to take a quick canter through the origin story of the modern factory.

A brief history of how we make things[9]

Here is some good news. With the benefit of hindsight we can look back at three hundred years of manufacturing history and see that things fall rather neatly into reasonably distinct phases, the most well known of these being the industrial revolution of the 1700s. Before this, most making was in the category of 'craft'. This describes production by hand in small workshops, using simple tools, and in low volumes. Then, thanks to a whole bunch of serendipitous innovations and interconnected social and economic factors, there was a transformation in the way we produce goods.

The shortest summary I can give you is this: there was a shift from person-based to machine-based production which led to substantial increases in both productivity (how many things could be made per worker) and the volume (the total amount) of things that could be made. These increases were so huge that we moved from just producing the things we needed to producing things for which demand needed to be created.[10]

The story of how that first revolution came about and led to subsequent revolutions is worth a few minutes of our time. It reveals something very important about why the manufacturing world looks the way it does today, and it also shows how our relationship with the manufacturing world as consumers has changed . . . and must continue to change.

The First Industrial Revolution

The period from the late 1700s to early 1800s seemed to be quite popular for revolutions. While America was separating itself from British rule and France was separating heads from the bodies of aristocrats, in the UK the revolution was more . . . well, industrial.

What became known as the First Industrial Revolution was the result of the convergence of several developments. Individually, each of these would have led to some improvements in how we make things; in combination, they were transformative. And one of the most important ideas came from a Frenchman whose ideas were so controversial that he had to be locked up for his own safety.

Honoré Blanc, a gunsmith from Avignon in southern France, put on a world-changing demonstration in 1785.[11] His aim was simple: to break the fixed mindset of the gun-making community. At that time, everyone who made guns knew that each component of a gun had to be made for a specific weapon. You'd make a rough part and then file and fiddle away to make the part fit that particular weapon. If something broke, you had to go and find an armourer who would mutter about your clumsiness and grudgingly make you a new bespoke part. It was not very efficient and certainly not practical in the middle of a battle as you needed to have a whole mobile blacksmithing operation accompanying troops to the front line.

What Blanc proposed was using various techniques for making parts of such consistent accuracy that any component would fit any weapon – the parts would be *interchangeable*. This was a brilliantly simple (and, with hindsight, blindingly obvious) idea that would dramatically speed up the process of manufacturing at scale and simplify the process of repair. Parts would be made using standardised measuring 'jigs' to ensure that every single one was made to the same dimensions. Boxes of spare parts could then be transported with the troops so that if anything broke, soldiers could just be given a replacement part that would fit any weapon. However, this idea was hated by gunsmiths who feared that the value of their skills would be eroded if all they were doing was making standard parts using foolproof fixtures and tool guides. As Simon Winchester notes in his book *Exactly*, there was such hostility targeted at Blanc and his workers that the French government moved them into a basement in the Château de Vincennes east of Paris to allow them to continue their work without being beaten up by disgruntled fellow gunsmiths.[12]

By mixing up components from different weapons and reassembling them without a hitch in front of an audience of sceptical generals and seething gunsmiths, Blanc had given a clear demonstration of the value of this approach. While the start of the bloodier type of revolution in France slowed further progress, one of the attendees at Blanc's demonstration was particularly taken by what he saw: the US Minister to France, Thomas Jefferson. Jefferson took these ideas back to the US where they were – after some rather interesting shenanigans[13] – widely adopted.*

* This approach led to what became known as the 'American System of Manufacturing' that underpinned the high-volume production not just of reliable weapons but soon of clocks, watches, sewing machines, flour milling machines, crop harvesters and much, much more.

Making reliably interchangeable parts required two things. First, you needed to be able to set the permissible limits of variation from the ideal product dimensions – what engineers refer to as *tolerance* – and assess whether each item met this standard. This required the development of metrology devices that could measure dimensions very precisely.[14] Secondly, you needed *machine tools* that could consistently produce parts to the desired tolerance.

'Machine tool' is a label given to any device that can consistently make dimensionally accurate parts by cutting, grinding, boring, shearing or deforming material. One of the first of these was made by John 'Iron-Mad' Wilkinson in the late 1700s for the precision-making of cannon barrels. Similar devices were then developed for making steam-engine cylinders, and then pretty much every manufactured part that required precision and consistency in high volumes. Machine tools allowed you not only to make high volumes of precision products for the customer but to make precision parts for machines making other things – they became 'a machine to make a machine'.[15] Precision machines could now be built in high volumes that allowed the reliable production of other goods, like textiles, clocks, agricultural tools and, of course, weapons.

But these machines needed power. Either the factory had to be located close to a river to capture the energy from the flowing water, or a new-fangled steam engine could be used to generate power wherever it was needed. The control of sources of energy and the use of machine tools to enable interchangeability moved manufacturing from a world of hand-based, low-volume, craft-style making to machine-based, high-volume production at a never-before-seen scale. We had entered the era of mass manufacturing.

But something was on the horizon that would crank production levels up by orders of magnitude. So great was this thing that it triggered the rather predictably named Second Industrial Revolution.

The Second Industrial Revolution

This new revolution was largely shaped by what was happening in the rapidly growing US economy of the late 1800s to early 1900s. Much of this was an acceleration and supersizing of things introduced during the first revolution in Europe, predominantly but not exclusively in the UK. These included the widespread adoption of those precision machine tools and the parallel advances in metrology,[16] as well as the massive scaling up of inexpensive ways of producing steel* and the shift from steam and water power to electricity.[17] A defining aspect of this US-focused phase of manufacturing evolution, however, arose from automotive entrepreneur Henry Ford's musings on what happens in an animal slaughterhouse.

From a manufacturing perspective, a slaughterhouse is a specific type of factory, and Ford was intrigued by one particular feature of how the more advanced abattoirs operated. A powered conveyor brought each carcass to a worker who then performed a particular task – cutting and removing – before the line carried the now lighter carcass to the next worker.[18] Having seen how efficient this process was, with each worker standing still and repeating the same activity as things passed by them, Ford applied it in reverse to the world of cars and, *voilà*, the assembly line was born.[19] Soon, thanks to the productivity gains realised from the widespread adoption of the assembly line, huge volumes of standardised products could be made.

But this triggered a new problem: how do you ensure that every single one of these thousands of products whizzing off the end of the assembly lines is being made to the required quality? You could end up rapidly assembling lots of things that, when inspected, weren't actually good enough. These would then require lots of what is

* Thanks to processes such as those developed by English inventor Henry Bessemer.

called *rework* – fixing and bodging things to remake them as they should have been made in the first place. This is a real waste of time and money. Factories needed to employ teams of inspectors to look out for defective products coming off these high-volume assembly lines. Squads of workers were needed to put right the errors spotted by the inspectors. If the defects were really bad, the products had to be scrapped. If only there was a way of ensuring that it was near impossible for a defective product to be made in the first place . . .

Luckily, there was, and its development was linked to solving the problem of working out who was living in Japan at the end of the Second World War.

In 1947, American engineer and statistician W. Edwards Deming was invited to assist the Japanese government with their national census. He clearly did a pretty good job as he was invited to stay on to teach statistical and quality control to Japanese manufacturers who were struggling to consistently make reliable products in high volumes.

To say that these manufacturing firms enthusiastically embraced his ideas doesn't capture the scale of Deming's impact on the Japanese economy. One particular company became so successful that they helped grow the Japanese economy from its state of near total destruction in 1945 to the world's second-biggest by the late 1960s. This company was Toyota, and the 'Toyota Production System' (TPS) developed by Taiichi Ohno and Eiji Toyoda – and cloned by many other firms – is what enabled the Japanese car industry to become so dominant in the 1970s and 1980s. By focusing relentlessly on efficiency of production, Japanese manufacturers were able to consistently produce cars that UK buyers, such as my parents, found amazing. These cars did things like starting on a cold morning, had windscreen wipers that *both* functioned, and were delivered on time with all the requisite parts attached.*

* That might not sound that impressive, but many UK car factories of that era were unable to consistently achieve such basic levels of quality.

The widespread use and visible success of TPS and all its unimaginatively named variants* spawned the broader concept of lean manufacturing, Ohno's own waste-minimising manufacturing philosophy encountered in Chapter 1.[20] And having products coming off a production line that require additional work to be done is clearly wasteful. You want things that are made right first time.

To achieve this, every single worker has to realise that there is no one at the end of the line who is going to correct any errors. Any problems must be fixed when and where they occur. To do this, workers on the assembly line must be given the power to stop the line as soon as they spot a problem.

When this idea was first proposed, it seemed heretical. Everyone knew that a smooth, uninterrupted flow of inputs being processed to a constant drumbeat to deliver outputs was essential. If you were working on an assembly line and made a mistake or saw something going wrong, you let it pass on to the next stage. Why? Because you knew it would be fixed by someone else later down the line. It was not your problem; admitting you'd made a mistake and stopping the line would likely just get you into trouble. But empowering and supporting shop-floor workers at every stage of the production line to be responsible for the quality of the things they were making was transformative. It is thanks to the widespread adoption of these ideas developed by Deming, Ohno and Toyoda that today we as consumers have access to such an extraordinary range of nearly defect-free, low-cost manufactured products.

So, by the 1960s, we had assembly-line-based factories that were implementing TPS-like approaches to allow mass production of a staggering variety of goods assembled from interchangeable parts. Then, along came another step change.

The robots arrived.

* E.g. Rolls-Royce Production System (RRPS), Volvo Production System (VPS), Bosch Production System (BPS) and many more.

Well, actually, while the arrival of industrial robots *was* significant, it wasn't the most important thing that was happening. It was the broader issue of increasing the amount of *automation*: getting technology to do things previously done by humans. In particular, it was about the automation of mental processes – such as data processing and analysis – rather than just physical processes. And the key technology that enabled automation and triggered manufacturing's next revolution was the development of the relatively low-cost electronic computer.

The Third Industrial Revolution

Electronic computers had been used by manufacturing firms since the late 1940s, but these early computers were huge, eye-wateringly expensive machines that only the largest of companies could afford to buy.* However, thanks to Nobel Prize-winning miniaturisation of the electronics that sit at the heart of all computers, by the end of the 1960s and into the 1970s, computers became orders of magnitude smaller and cheaper. As a result, manufacturing firms of all sizes began to access computing power to help them run their factories. Data from factory operations and market analysis could be processed more quickly and accurately, allowing manufacturing to become more efficient and granting firms the ability to respond to changes in customer demands more effectively.

Then, in the 1990s, the world of computing got an injection of adrenaline: engineers worked out how to join up all these small computers into a giant *inter*connected *net*work, which got shortened to . . . the internet. This pushed things into a whole new league. Manufacturers now had the ability to connect to data from

* And I was delighted to learn that one of the major players in the early development and use of computers in industry was a UK cake manufacturing company. We'll come back to the extraordinary story of how that happened in Chapter 6.

a much wider range of sources, and they could access, process and share data within and between their own factories as well as those of their suppliers. And as the internet became widely available to us as consumers, manufacturers realised that this could dramatically improve their understanding of our needs and wants, enhancing their ability to deliver the products we required or craved. Thanks to internet technologies, manufacturers were able to connect activities along the whole make-move-consume journey.

In this Third Industrial Revolution, we had factories that were producing (nearly) defect-free, reasonably priced products in staggering volumes and variety. Computers and robots had made the input > process > output cycle within these factories amazingly efficient, able to respond at speed to changes in customer demands.

The problem with industrial revolutions compared to, say, political revolutions is that there isn't a moment you can point to and say, *Ah, that's when the mob lit their torches and stormed parliament and the revolution started.* The industrial versions seem to be more evolutionary than revolutionary, and sometimes they are only evident when well under way or when viewed through the lens of hindsight.

So, where are we now?

The Fourth Industrial Revolution (a work in progress)

Not everyone has signed up to the idea that we are in the midst of the Fourth Industrial Revolution. Many take the view that what we are seeing is just the ongoing – albeit supercharged – evolution of version three. Whichever view you take, the way we make things is definitely changing.

Some of the drivers of this current change are technological: the phenomenal increases in available data and low-cost computational power enabling the power of AI to be unleashed on almost every aspect of life. Some are geopolitical: the changing balance of power triggering brinkmanship and actual conflicts that disrupt supply chains. Others are more sociological: our globalised, hyper-connected world allowing pandemics to spread rapidly.

But the most significant driver of current changes to the manufacturing world is sustainability. We now have a crystal-clear realisation that the way we make things has a direct and significant impact on the ability of human life to continue on earth. Untold levels of environmental damage have been caused by centuries of fossil-fuel-fuelled, extraction-focused industrial activity, and our economic model of consumerism relies on ever increasing levels of making and selling things, much of which is disposed of long before its useful life is over, creating mountains of waste.*

But what we are seeing now is, I think, a really positive start towards a very different, more sustainable approach to manufacturing. Whether we want to call that a new industrial revolution or not isn't really important. What is important is that we are changing the way we make things not only to cause less damage but actually to start *repairing* the damage already done. And this new model will not only be better for the planet, but also be more resilient to disruptions and provide better opportunities for those who work within it. How this is happening is something we will focus on in the second half of this book.

We now have the first essential tool for understanding why the manufacturing world is the way it is. I've given an overview of the basic ways things can be configured to deal with the 'make' part of the manufacturing world. We've looked at how what happens in all factories can be framed in seven words: they take some *inputs*

* One of the most infamous being one made up of tens of thousands of tonnes of 'fast fashion' clothing dumped in Chile's Atacama Desert https://www.wired.com/story/fashion-disposal-environment/.

and do some kind of *processing* (*people* follow some *method* using *machines* and *materials*) to convert them into more valuable *outputs*. We have seen that there is a strong relationship between product volume and product variety, and how that relationship dictates what type of factory is needed to make the product. And we have also shown how we, as consumers, have changed our relationship with the manufacturing world as we moved from a simple, often local approach of craft-based, make-what-you-need manufacturing to the global production of goods at mind-boggling levels of scale, variety and quality. We are now well placed to move on from the 'make' part of the story to the 'move' part.

Almost everything we've discussed here – apart from my trip out to buy some cocoa – has been focused on things happening *inside* individual factories. We've deliberately focused on the process bit – the methods, machines, materials and people – that sits between the inputs and the outputs of the factory.

For the next part of our exploration of the manufacturing world we need to head *outside* the factory, where things are going to get a little bit more complicated. But don't worry: by joining me in unpacking a new bicycle from its box, climbing a dizzyingly high crane at a container port and investigating how ice cream remains cold once it leaves the factory, all will be made clear.

3: Move

Why manufacturing is all about journeys

I swore as I slammed on the brakes. The truck ahead had pulled out of its lane to overtake a fractionally slower one in its path. Fuming and muttering as it completed its glacial overtake and finally pulled back over to the left lane, I swept past with a childish flourish of engine-revving and gear-changing . . . before sighing as I slowed down for yet another reshuffling of the order of the other trucks ahead. As the dual carriageway crested the incline and gently dipped into a wide valley, the increasingly slower lanes of cars, vans and trucks stretched right to the brow of the next hill.

To distract myself from the boredom of being in what soon became stationary traffic, I started looking in more detail at the trucks. Some displayed familiar names, others unfamiliar ones, a few blandly anonymous. Most had UK numberplates, but several were from the European mainland. I started thinking that within each shipping container, inside each box or curtain-side trailer, sloshing in each tanker or fixed atop flatbed trailers would be the materials or components or finished versions of a thousand and one different products. Each vehicle was carrying something whose total journey might range from a couple of miles to a multi-thousand-mile odyssey. Some would be bringing raw materials to factories, others moving part-finished items between factories, and others bringing final products to distribution centres and shops to be delivered or collected by consumers like you and me.

I wasn't just seeing a traffic jam. I was looking at a snapshot of links in multiple manufacturing supply chains.

As I sat re-caffeinating myself at a service station a short while later, I thought a bit more about what I'd just seen – and what we all see on the roads each day. The manufacturing world relies on a lot of moving stuff around. *A lot.* And it made me think: if an alien came down to earth and looked at how we make all the things needed to live our lives, I reckon they'd be quite puzzled and ask, *Why do earthlings spend so much time and effort moving stuff around the planet to make the things they want?*

By the end of this chapter you'll be well prepared for the day when our new alien overlords summon you to provide an answer – but I don't think they are going to be very impressed with your response.

The logic of logistics

In the world of manufacturing, all this moving-things-around falls under the heading of *logistics*, defined by my dictionary as 'the careful organization of a complicated activity so that it happens in a successful and effective way'.[1]

Every place where things are made will be managing logistics in some shape or form, whether it is me wandering about the kitchen picking things out of the cupboard, conveyors feeding machines with electronic components to assemble phones, or automated guided vehicles moving car parts to workers on production lines. As the previous chapter showed, when such systems work as intended, everything flows smoothly through the factory from goods-in to dispatch. But this won't happen if the right materials aren't delivered to the factory and the finished products aren't packed up and shipped off to the appropriate customer on time.

Factory managers know that achieving the nirvana of a perfectly flowing manufacturing operation requires the synchronisation of activities not only within but also beyond the factory gates. And while managers have control over things within their own factory – the *internal* logistics of all the activities discussed in the previous chapter – managing external suppliers and customers can be more challenging. It's the same as controlling what happens in your own kitchen compared to the relationship you have with your local or online supermarket.

To understand how this synchronisation happens, we need to take a look at what happens beyond the factory gates and explore the rather predictably named world of *external logistics*.

The twenty-thousand-kilometre journey of a bicycle

While I was sitting in a meeting, my phone buzzed to alert me to an important message: my new bike had been delivered to my home. Just as it had been when I was five years old, this was a day of great excitement to me as a fifty-five-year-old. Since my trusty old bike* had finally given up the ghost, I was forced to buy a new one, one more fitting for my age and needs. No more carbon fibre and shoe-clippy pedals for me. Time for mudguards and a sensible upright seating position.

A large brown cardboard box, slightly damp with some dents and tears, was leaning against the garage wall when I got home from work. Too excited to go into the house, I tore open one end of the box and peered inside. There, packed with origami-like neatness, I could make out a complete bike. One jarring feature was the handlebar positioned parallel rather than perpendicular to the frame. It was like seeing someone with a badly broken arm or leg: it just shouldn't be pointed that way. To rectify this as quickly as possible, I put one foot on the box, pulled out the too-shiny red bike and leaned it gingerly against the rough brick wall. Squinting back into the depths of the packaging, I saw a small white cardboard box nestled in the far end. Tipping the package up, the little white box slid out. Neatly arranged inside were all the tools needed to complete the final assembly. Five minutes later, my bike was ready for the road.

* The one I had ridden since becoming – to the horror of those close to me – a MAMIL (Middle-Aged Man In Lycra).

As I did a final check to make sure everything looked and felt right, I noticed the small black sticker at the base of the frame. In white letters were written three words common to about a third of the world's manufactured products: 'Made in China'.[2]

That unboxing, handlebar-straightening and assorted nut-tightening marked the end of this bike's journey of over twenty thousand kilometres to start its working life speeding me to and from work. And that journey provides a pretty good illustration of all the key features of the extraordinarily sophisticated, tightly co-ordinated – yet slightly crazy – world of external logistics.

My bike is a bog-standard commuter bike and was probably made a couple of months before it arrived at my front door. Like all bikes, mine has a frame, two tyred wheels, a pedal/chain/gear arrangement, handlebars, brakes and a seat; and as with most bikes of this type, the frame is made from a set of welded aluminium tubes. The spoked wheel rims are pressed from aluminium, and the gears and brakes have been bought by the bike company pre-assembled from a specialised manufacturer (in this case the Shimano Corporation of Japan). Various other bits – mudguards, lights, seat – are bought in from other dedicated suppliers.

My particular bike was made 'to stock' for a bike retailer. That means it was made before a final customer (me) had ordered it. Based on their analysis of the market, the retailer was pretty sure that someone like me would want a bike like this. They had ordered a bunch of them so that when I placed my order it could be dispatched straight from a warehouse in the UK to my house within a couple of days.

Though it has the brand of a US-based company, in common with most high-volume bike manufacturers that company gets many of its bikes manufactured in China.[3] In one of many enormous factories located across the industrial east side of the country, it takes just a few hours to assemble the hundred or so parts needed to make mine (one of over seventy million produced each year in China – about three-quarters of the total produced in the world).[4] While some components are made within the factory, others have been shipped in from suppliers' factories located mostly elsewhere in China – having been formed and welded, the metal bike frame is loaded onto a conveyor system to move it around the factory and have all the other components added to create a working bike.

At the output end of the factory, the bikes are slotted into the bike-sized cardboard boxes with appropriate protective internal packaging known as 'dunnage' before being neatly slotted into shipping containers waiting at the dispatch end of the factory. Each container is then taken by train or truck to a shipping port. From there, my bike-containing container is loaded onto a ship to complete the longest step of its journey,[5] one that takes over forty days and covers around twenty thousand kilometres.[6] Locked away in the dark in one of many thousands of containers stacked below and above the deck, my bike is oblivious to its passage across the South China Sea, through the Strait of Malacca, and over the Andaman Sea, Bay of Bengal, Indian Ocean, Laccadive Sea, Arabian Sea, Gulf of Aden, Red Sea and Gulf of Suez. It then squeezes through the Suez Canal (assuming it's not been blocked again) before emerging into the Mediterranean Sea, passing through the Strait of Gibraltar, edging along the eastern fringe of the North Atlantic Ocean, crossing the Bay of Biscay and heading up the English Channel before finding its berth in the UK's busiest container port, rather improbably located at the edge of the sleepy Suffolk town of Felixstowe.

Attention! Fasten seatbelt before driving crane.

That warning is stuck above the cab window as I lean forward from a remarkably comfy chair high above the quay at Felixstowe docks. I peer down between my safety-booted feet through the metal mesh floor. The chair's armrests each have black plastic video-game-like joystick controllers. The right-hand side also has an A4-size control screen and a reassuringly bright red 'STOP' button. The crane's boom stretches out ahead of me over the hundreds of multicoloured containers stacked on a pale blue vessel owned by the A. P. Møller-Maersk shipping company. Hanging on cables from the crane's dark-blue boom is the yellow 'spreader', the device that is clamped onto the top of containers – each one of which could weigh up to fifty tonnes – to move them to and fro between ship and quayside.

My hands sweat in my safety gloves as I reach tentatively for the controls.

'It's OK,' says Dave, my host. 'You can't do any harm – we switched everything off on this one.'

I hadn't really thought that Dave was going to let me drive this £12 million machine, but I still felt a slight tinge of disappointment.

A few minutes later, standing on a breezy walkway at the back of the crane, I try to take in the bigger picture. In the gentle spring sunshine, stretched in front and to either side of me are thousands upon thousands of containers – from this distance they look like ridged, elongated Lego bricks. Amongst the white of those labelled 'Maersk' and the dull yellow of 'MSC', the appropriately bright-green containers from the 'Evergreen' shipping company stand out.

If my eyes struggle to process the scale of what I am seeing, my ears fare no better with the words Dave is sharing. The crane he was letting me pretend to operate is one of thirty-three stretched along nearly four kilometres of quayside. Each year, the port handles four million TEU containers, and the port is a parking area for up to 150,000 such containers. Even more mind-boggling than this is the fact that even though Felixstowe is the busiest container port in the UK, accounting for nearly 30 per cent of all containers coming in and out of the country,[7] it doesn't even make it onto the list of the fifty busiest container ports in the world.*

But back to containers. If you sometimes struggle to remember where you parked your car in a multi-storey car park, imagine the challenge of knowing not only where each individual container is among the other 149,999 but also exactly when it is going to be collected and by whom. You also have to know what is inside every single container, to ensure that dangerous or perishable contents (temperature-controlled in what are known as 'reefer' containers that have their own chilling units) are placed in the right location to ensure safety or access to power. There is some very complex software at work to play this 3D, hyper-scale, physical game of Tetris.

So, what do all those thousands of containers contain? A trawl through ships' manifests later reveals that those containers spread before me are likely to include (in no particular order) bananas, bottles of wine, computers, toys, tyres, electric heaters, silver, car parts, furniture and, of course, bicycles.

Having spent some time stacked in its precisely chosen spot in that multicoloured, multi-storeyed container park in the Port of Felixstowe, my bike's container will be extracted via a rubber-tyred gantry crane (one that looks a bit like a giant four-legged stool on wheels that hoists up containers with cords suspended from beneath the 'seat') before being carefully positioned over and locked onto the back of a waiting articulated truck. The truck then heads out, stopping briefly to have its paperwork checked at the gatehouse, and turns west down the A14 dual carriageway to a distribution centre in the middle of the UK.

* That list is topped by Shanghai which shifts *fifteen times* more containers than Felixstowe, that's more than the top five US ports combined. https://en.wikipedia.org/wiki/List_of_busiest_container_ports. And as an aside, big though the numbers relating to shipping containers are, containerised freight only makes up around 14 per cent of the nearly half a billion tonnes of goods that arrive at and leave UK seaports each year.

A few hundred miles and a coffee break later, the truck will pull into an industrial park to offload at the distribution centre. There, the container is checked in and assigned to a specific loading dock. It is lifted off the truck using a specialised vehicle called a container handler or a side loader and placed on a designated loading dock. Depending on whether the goods in the container arrive on the floor of the container or on pallets, the unloading process involves a synchronised combination of manual labour, forklifts, pallet jacks or conveyor systems. Once the cargo is unloaded, it is sorted according to its destination or specific requirements and placed in a very specific position within the vast, aircraft-hangar-sized building.

There, my bike waited patiently in its box until the online order came in from me sitting on my sofa via a few clicks on my laptop. My bike was then dispatched via one of tens of thousands of smaller delivery trucks, arriving at my house a day or so later.

More complex products = more complex logistics

As we now know, even a simple product such as toilet roll will have gone on an extraordinary set of journeys to reach us consumers. With more complicated products, with lots more sophisticated components, the logistics get more involved. The bike example shows an assembled product of medium-ish complexity – the making of it required the bringing together of around one hundred components (if you include every nut and bolt). Now just imagine the complexity of the logistics needed to make and deliver products with thousands or millions of components, be it a phone or an airliner. How do the external logistics for really complicated manufactured products differ from the one for my new bike?

For a phone, one difference is the number of times parts or components are shipped to and from locations around the world. Why? Because the very specific skills required to make the different systems within today's ludicrously sophisticated phones don't all sit neatly in one place; no single company possesses the huge range of skills needed to make the exquisitely complex communicating, sensing, processing and actuating subsystems that make up a typical modern phone. As a result, part-finished materials, components and subsystems have to be shipped between companies all around the globe to have some particularly exacting process completed or some fiddly component added.

What does all this add up to? David Humes in *Door to Door* estimates that the total distance travelled by an iPhone and its components before it reaches you is at least 250,000 kilometres – the equivalent of over six times round the world.[8] That's over twelve times farther than my bike travelled to get to me . . . and that already seemed a ludicrously long distance.

For an airliner, the complexity of logistics is one result of the sheer number of component parts. Every single one must be perfect, and that perfection must be traceable. Why? Because in the case of any malfunction, whether tray-table-clip minor or falling-out-of-the-sky serious, the exact point along the supply chain where any production defect occurred must be identified to ensure it cannot happen twice.

Another difference between my bike and an airliner is the size of some of those parts. Take a typical wide-bodied, long-haul plane such as an Airbus A350 as an example. It has two engines weighing over seven tonnes each; these are five metres long and have three-metre-diameter fans (the main working part you see at the front of the engine). This aircraft has thirty-two-metre-long wings and more than nineteen tonnes of landing-gear assemblies (the brakes, wheels, tyres and shock absorbers).

You can imagine the internal logistical challenges faced by the factory manager in moving all these parts around to make sure they are stuck together in the right order. But, as with the bike and phone, many of these components are actually made elsewhere and then shipped, in this case to Toulouse in southern France, for final assembly.

Geopolitical logistics

Having bits made elsewhere is very common in manufacturing, but in the case of Airbus there is an additional overlay of geopolitics that influences the design of their external logistics.

Airbus was formed as a consortium of European aircraft manufacturers in the 1970s when it was realised that, given the massive cost of developing the hugely complex thing that is a modern long-range commercial airliner, none of them was going to be able to compete with the mighty US-based Boeing corporation. That was all very sensible, though it did lead to some headaches for those responsible for logistics.

Sharing the work of building aircraft amongst the various organisations that form the Airbus consortium means that major parts of the aircraft are made in different factories, many of which are far away from each other and from the main assembly plant in southern France. As a result, some of the big components of the A350 are made in different parts of the world and then moved by ship or specially adapted aircraft* to be assembled in Toulouse. What sort of things are moved? The engines could be made and shipped from the US (if the customer wants GE engines) or the UK (if the, er, Rolls-Royce of aeroengines is requested). That makes sense as making jet engines is a very specialised activity, and you want to buy these from specialist suppliers rather than trying to make one yourself.

Given the size of aircraft wings you might think that it would be sensible to make these in or close to the main Airbus factory in Toulouse, avoiding the challenge of shifting something coming in at over twenty tonnes, thirty metres long and six metres wide. The wings are actually made nearly seven hundred miles away in Wales and flown in one of those specialist aircraft all the way to

* These aircraft are cartoonishly freaky. Just type 'airbus beluga' into a search app and you'll see what I mean.

Bremen in Germany, where the 'high lift' systems – the flaps and slats that allow the plane to fly at lower speeds – are installed before being transported to southern France for attachment to the rest of the aircraft.

And the wings are not the only part that does this. The back end of the fuselage and vertical tailplane is made in northern Germany; the horizontal tailplane is made in central Spain; and the landing gear is made in either the UK or Canada, depending on the specific flavour of A350 being made.

This geographically dispersed co-ordination is managed with astonishing levels of precision.[9] As one result of doing that so well, there is a thriving aerospace industry in Europe employing over 400,000 people. And, should you or I need to, we are able fly on reliable and safe aircraft, manufactured at a cost that allows airlines to run their business at a profit.

Supply chains sometimes need to chill

However, it is not just the quantity, sophistication or size of components that can make the management of external logistics a major headache for manufacturing firms. What if the raw materials, components or products themselves need to be kept at a consistent temperature throughout the journey from raw material to production to consumer?

Moving products like these requires the design and operation of a special type of supply chain known as the *cold chain*.

The cold chain is a label given to the system of moving and storing temperature-sensitive things. It is a very specialised part of the world of logistics and even has its own dedicated industry associations, such as the UK's Cold Chain Federation.

For those shipping foodstuffs like ice cream between factory and consumer, their product must be consistently maintained below −18°C. From factory to truck, truck to distribution centre, distribution centre to another truck, and truck to supermarket: the 'cold chain' cannot be broken. If at any stage the temperature has risen above −18°C, even if just for a moment, ice crystals might have formed that could ruin the customers' pleasure and drive them to angry, brand-damaging rants on social media.

For other products, allowing the product to get a bit warm in transit is likely to lead to more serious, health-impacting problems for the trusting consumer – think seafood, or even vaccines. Organisations like the Cold Chain Federation can ensure that logistics firms are fully compliant with all the latest legal and safety requirements (not just for their products but also to avoid over-chilling their staff, too).[10]

There is of course a big challenge facing those who manage cold chain logistics: how can they reduce the environmental impact of running all those temperature-controlled trucks and distribution centres?

In addition to the obvious things such as switching to renewable energy sources, food manufacturers themselves are finding ways to make their precious produce need less cooling during transit. For example, ice cream manufacturer Unilever recently found a way to make ice cream that remains stable at −12°C rather than −18°C. That doesn't sound too earth shattering, but this means that the amount of energy consumed by ice cream freezers was reduced by 25 per cent. And in an altruistic gesture, Unilever made the patent for this new ice cream recipe freely available to others.[11]

Those three examples – the phone, the plane and ice cream – coupled with the description of the (comparatively) simple bicycle have, I hope, given you a sense of the complexity and skill that sits under the rather uninspiring label of 'external logistics'. Right now, literally millions of people spread across every country on earth are involved in the planning and co-ordinating, shipping and flying, maintaining and improving, packing and sorting needed to ensure that every aspect of this vast 'system-of-systems' operates (mostly) flawlessly to keep all the factories working and us as consumers satisfied.

The fact that these vast external logistics systems work so well almost 100 per cent of the time is one of the most miraculous features of the manufacturing world.

So, you are almost ready to respond to that question posed by the visiting aliens. Hopefully you now have a picture in your head of how things are moved around, and why we do this. But there is one more issue we need to explore before you collect your notes and wait nervously to be summoned to deliver your answer: have we always moved stuff around this much?

How we shrunk and expanded different parts of the manufacturing world

In 1997, my PhD was 'nearly finished'* – but I had run out of money. To address this problem I needed to find a part-time job, and my supervisor had heard that one of her colleagues was looking for a research assistant. The role sounded quite glamorous when explained to me over a cup of instant coffee in the basement of the engineering department on a damp Friday evening – I was told I could join a project investigating a phenomenon that was being called 'international manufacturing mobility'. In response to my evidently gormless expression when told this, my soon-to-be boss Mike patiently gave some more detail.

* A vague descriptor commonly used by PhD students to describe a wide range of states of thesis completeness from 'just doing final checking of the spelling' to 'nearly started typing some words'.

My role would be to visit factories to capture detailed information about how a product was made in one place so that the same product could be made in another place. These places were hinted to be China, the US and Mexico, which sounded very exciting given the implied involvement of international travel, and so I eagerly signed on the dotted line. A few days later, as I drove my hire car through a gap in a rusty chain-linked fence towards a part-demolished factory complex in a decaying corner of the UK's manufacturing heartland, I started to suspect the role might have been slightly mis-sold.

Reflecting on this experience, I now realise I was playing a microscopically small part in one phase of a massive ongoing transformation of the world of manufacturing that started many years ago.

From DIY to outsourcing

Henry Ford transformed the manufacturing of cars. He became globally famous for the way in which he organised production to the point where if someone said 'assembly line' it invariably triggered the response: *Oh, like Henry Ford and the Model T.** But there was something else that Ford (and many others) did as they developed large-scale manufacturing operations at the start of the twentieth century.

To ensure that the Ford Motor Company always had access to all the things it needed, his firm did almost everything itself. Ford had its own steel mills to produce the steel for the car bodies and a glass-making plant to produce the windscreens and windows, and it even owned forests to ensure a reliable supply of wood for other components. What's more, beyond the making of parts, Ford also used to run its own fleet of delivery trucks, canteens

* So strong was the link that this approach became referred to by many as 'Fordism'.

to feed its workers, in-house customer services, and, well, almost everything.

This approach is known as *vertical integration*: all the materials and process steps needed along the chain of activities from idea to product are owned by the firm making the product. Vertical integration allows a manufacturing firm to maintain control over everything and avoid all that tedious faffing about with suppliers, arguing over costs and getting stressed when parts don't show up. With vertical integration, everything is within your control.

However, there is a downside.

What happens when things change? Sticking with the early days of Ford, imagine being faced with this situation. The early cars were started using a crank handle. This provided drivers with an exciting workout each time they wanted to go for a drive, and a hint of jeopardy as there was always the risk of a dislocated shoulder or broken thumb. When someone invents the electric starter motor, you realise that your customers might quite like this, but you don't know how to make electric starter motors. So, with a vertical integration approach, you might summon your engineers and tell them to work out how to make a starter motor and set up a production line. Or you could tell your finance people to go out and find a firm that makes starter motors and buy enough shares to make that other firm part of the Ford Motor Company. Or you could just ask one of the starter-motor companies to supply you with some of these new-fangled devices.

That last option is known as *outsourcing*, and, over the years, the widespread adoption of this approach has transformed the way we make things. Today, if you look at Ford or any automotive company, most of the parts that make up each car are not made by them.[12] They will be made by a network of hundreds – in some cases, thousands – of suppliers, each one a specialist at making some particular type of component.

Reducing the cost of doing things at a distance

Outsourcing is one of the two major things that have transformed external logistics. The second is globalisation. We can also look to the past to explain this phenomenon, this time to the 1950s, when a US trucking entrepreneur, Malcolm McLean, was trying to start a revolution in shipping.

McLean transformed the transportation industry by implementing an idea of such elegant logic it makes you wonder why it hadn't happened before: instead of the faff of loading individual items on and off ships, trains and trucks, why not put all the things in big metal boxes and move the boxes around?

McLean, however, understood that to make this superficially simple idea work would require standardisation and collaboration. You needed to convince everyone in the freight industry to agree on a common design for these containers so that they could seamlessly be transferred between ships, trucks and railway cars; shipbuilders would have to design ships that would take container-shaped cargo; port owners would require new cranes for handling containers; and ports would have to be redesigned to optimise the flow of containers into and out of those ports.[13]

Once you've successfully done all that, though, you have what is called an *intermodal freight transport system*. And then it gets much easier – and much, much cheaper* – to move lots of things over long distances. And if you can move things around cheaply, making things in factories far away (where, for example, the cost of labour is much lower) or shipping things to customers far away becomes a cost-effective option.

* To give a sense of the reduction, Marc Levinson in *The Box* (2016) reports that McLean's company claimed the cost of loading a non-containerised ship in 1956 was $5.83 per tonne. Using his first attempts at containerisation, the cost was reduced to $0.16 per tonne, i.e. a cost reduction of 97 per cent. With numbers like that, it was pretty much instantly game over for non-containerised shipping.

Manufacturing could now be distributed and managed on a global scale.

As a result of outsourcing and globalisation, the 'footprint' of many manufacturing operations spread out across the globe, and this evolution has had some long-term major consequences for how different nations create wealth to support their economies.

How some countries forgot how to make things

Major manufacturing countries used to make products domestically and ship them overseas. Then factories started being moved to locations where labour cost less or workers had specific skills. If transport costs were high, it could become more efficient to make things 'closer to market'. And sometimes the government of the country in which you wanted to sell would demand that you make your products locally, using local suppliers. Why? Because this would allow that country to strengthen its own economy by developing local manufacturing infrastructure.

More recently, many large firms don't think so much about what factories they need but more about having access to what are rather grandly called *production capabilities*: the appropriate bundles of skilled people and machines needed to make a particular thing. With this mindset, a manufacturer removes the brain ache of running factories themselves and instead can focus on conducting an orchestra of other firms with the appropriate capabilities – wherever they might be located – who can make and deliver their products.

In the middle of all these seismic transformations, production capabilities were scattered all around the world. To facilitate this, someone had to understand what was going on in one factory and explain to someone else on the other side of the world how to make that same thing. That person had to visit factories, many of which had been in operation for decades at the heart of thriving communities, but were, in some cases, about to be shut down. And when that person arrived at the factory, their job was to interview people who had worked in that factory for many years and get them to explain what they did so that someone in another country could do their job instead, putting them out of work.

I was one of those people.

As I sat in that car park next to the part-demolished factory on a damp November morning in 1997, mustering up the courage to interview some highly skilled and understandably hostile factory workers, I began to rethink my life choices.

The nonsense of post-industrialisation

The end of the last millennium was an extraordinary period of change in manufacturing, and one that some economists saw as natural and healthy economic evolution. These particular economists raved about how natural and sensible it was that some nations had become 'post-industrial' economies. What did that mean?

Economist Ha-Joon Chang talks about how, through the 1990s, it was perceived as being much better for low-wage economies like China to engage in manufacturing while the richer countries focused on 'high-end services' like finance, IT and management consultancy.[14] Nations such as Switzerland and Singapore were touted as the exemplars of how an economy could thrive in such a post-industrial mode – but, argues Chang, this is just plain wrong.

In fact, the most industrialised economy in the world with the largest amount of manufacturing output per person is . . . Switzerland.* And in second place in the industrialised economy

* And, no, it is not all chocolate and expensive watches. The Swiss also have significant strengths in precision engineering equipment, industrial chemicals, pharmaceuticals and medical technologies.

league table? Singapore.*

Irrespective of this, having a post-industrial economy seemed like a great idea for those running the UK and US in the 1990s. These nations shifted away from 'traditional' manufacturing, and government policies were focused on building on strength in financial services along with a broader push towards what was known as the 'knowledge economy'. One key part of the knowledge economy is that it relies on the generation of ideas that can be protected and licensed to others to do something useful. You don't need to make things – instead you can get others to make things based on your ideas, but they pay you each time they do so. When this works, it can be really powerful.

A great example of a firm that does this is ARM, which designs the microprocessors that sit at the heart of 95 per cent of all our mobile devices. Yet ARM hasn't actually made any of the over 250 billion chips based on their designs – other companies pay them for the right to make those chips. Every chip that gets manufactured – mostly in Taiwan, as we shall see later – delivers revenue back to ARM. As a result, ARM was valued at over $150 billion in 2024.

But you can't build a whole strong economy based only on ARMs, investment banks, and management consultants. As Chang puts it, '. . . the ability to produce manufactured goods competitively remains the most important determinant of a country's living standards'.[15]

Having now taken a bird's eye view of global manufacturing logistics, we can understand with greater clarity what we mean when we say that manufacturing is complex ('involving a lot of different

* What is also interesting about successful economies like those two is that they have strong manufacturing *and* strong service sectors.

but related parts'). It requires the synchronisation of the movement of lots of those 'different but related parts' both inside the factory and beyond. We first looked at the movement of things within the factory, and we have now seen how things beyond the factory need to be synchronised to allow the smooth flow of manufacturing activities. As centuries have passed, the world of manufacturing has become ever more sophisticated. Economies have advanced thanks to technological progress (from power generation to information processing to communications to transportation) and social change (quality of life has improved and we now demand more things) – and the interplay between these two factors. Manufacturing systems have evolved to a point where they are heavily reliant on the seamless, rapid and low-cost movement of billions of objects around the globe.

However, the development and running of this mind-bogglingly intricate and interconnected logistics system has led to some rather spectacular unintended consequences.

When fragile supply chains snap

On 23 March 2021, the *Ever Given*, a 200,000-tonne, four hundred-metre cargo ship loaded with over eighteen thousand TEU shipping containers, managed to get itself wedged across the Suez Canal. This is an alarmingly narrow stretch of waterway through which move around 30 per cent of all shipping containers, representing 12 per cent of total global trade – that's about $1 trillion worth of goods.[16] This canal is so narrow – or rather modern ships are so large – that it can only allow ships to travel in one direction at any time. It took a week of frantic effort to re-float the *Ever Given* and unblock the canal, during which it was estimated that around $10 billion of trade was held up *per day*.[17] With this canal blocked, ships travelling between Europe and Asia had to add over five thousand kilometres and around ten days to their journeys by going all the way round the southern tip of Africa.

The *Ever Given* incident revealed how delicate our global logistics network had become. One person making one decision on the bridge of one ship can cause a major section of this network to instantly fail.

Our remarkable intermodal global manufacturing system can sometimes be a double-edged sword. We as consumers get the ability to buy a vast range of high-quality, low-cost goods, but the *Ever Given* example shows what can happen when things go wrong.[18] Now, you might be thinking that adding ten days to the delivery time of a new TV or some headphones isn't so bad. But what if the ships that have been delayed were transporting urgent medical supplies? Or perhaps components for a major engineering project, without which it cannot continue?

Rather worryingly, however, fragility is just one of the major problems with this system.

The true costs of making things far away

Another unintended consequence of running this supposedly super-efficient, hyper-lean system for moving things around is that, ironically, it is remarkably wasteful. Taking one seemingly ludicrous example, thanks to the low cost of transportation, under certain economic conditions, it can be economically advantageous to have fish caught off the Scottish coast sent thousands of miles to China for processing and then shipped back for sale in UK supermarkets.[19]

While such decisions might make economic sense, they ignore something that economists call *externalities*. An externality is something you do that affects someone or something else. The pollution caused by shipping frozen fish between Scotland and China would be an example of a negative externality.* It's not a problem for the fish processing company or for UK consumers, for whom costs are reduced, but the shipping generates pollution, which damages our environment, harming everyone else in the world who does not benefit from this cost reduction. The daily movement of raw materials, components and finished products of many types around the world generates megatonnes of such pollution.

This amazingly cost-effective system therefore actually contravenes one of the basic tenets of lean manufacturing: waste should be avoided at all costs – and that includes externalities.

So what can be done to reduce the environmental impact of all this moving things around? There are broadly two ways forward: reduce the distance you move things; and, if you have to move things, transport them in a cleaner way.

The positive impacts of shorter supply chains

The 'move less' option means cutting down the use of long supply chains, i.e. making things closer to where they are going to be needed. This sounds simple but, as we saw earlier, there are lots of good reasons why particular things are made in specific places.

Shifting patterns of geopolitics can create high levels of uncertainty for manufacturers. Firms get nervous when they realise that critical parts of their supply chains are overly concentrated in countries where things, for various reasons, are becoming 'complicated'. Uncertainty requires firms to consider relative levels of

* Externalities can be positive or negative. For example, if I buy a bicycle so that I can commute to work instead of driving, that delivers a positive externality: I have made a modest contribution to lowering the level of traffic congestion during rush hour. But I have also created a negative externality. The bike was manufactured in China and shipping it from there to Cambridge generated tonnes of CO_2 from the burning of the highly polluting bunker fuel powering the container ship plus the diesel trucks that carried it on its twenty-thousand-kilometre journey.

risks, and this may lead to the decision to relocate factories or change suppliers to more stable locations.

Another reason for reconsidering the impact of outsourcing and globalisation is something noted by Harvard professors Gary Pisano and Willy Shih while looking at the US economy back in 2009: 'today's low-value manufacturing operations hold the seeds of tomorrow's innovative new products'.[20]

Linking back to the earlier observations from economist Ha-Joon Chang, while the idea of the knowledge economy is all very well, there are real advantages in being able to make the things you've invented. To do that, it helps to have close links between those who are coming up with the latest ideas and those who are going to be making the new products, but if you've outsourced your manufacturing to other companies in far-away countries, you will slowly but surely lose your ability to make things. Once lost, this ability is very hard to regain.

The national ability to make things is linked to something labelled *industrial commons*. This draws upon the idea of 'common land' in rural communities that provides benefit to the whole community – if there is grassland on which everyone can graze their animals, all will benefit. In a similar way, having a network of skilled manufacturers and suppliers close by benefits everyone. If the skilled people and capable suppliers have found their services are no longer in demand as everything is made in a country far, far away, they'll go and do something else, and slowly but surely the country's industrial commons will face the equivalent of the grass on the village commons dying out and being used as a car park.

The words of Pisano and Shih led to much nodding agreement but not a great deal of action. When asked whether they were concerned about the loss of industrial commons, many politicians just shrugged and said something along the lines of *Well, that's globalisation for you. But look how cheap all those electronic gizmos now are!*

Right up to the financial crash of the late 2000s, they were adding, *and look how strong and stable our financial services sector is!*

Rudely awakened by that global financial crash, conflicts and a pandemic, by 2023 these same politicians were shouting about the importance of supply chain resilience and the need to 'reshore'* manufacturing – surveys recently showed that 70–80 per cent of US and European manufacturing firms intended to bring more production home or closer to home.[21]

As we've seen, there are one or two practical challenges to moving all the operations of one factory to a new location. Substantial time, effort and money will have been invested into making a factory work particularly well in one place, and there is a whole gamut of costs and risks that need to be dealt with when you attempt to move something as complicated as a whole factory's operations.[22] This is not something that any firm will undertake lightly.

And, as life is always more complex once you scratch away the surface, there's more than just one flavour of ways to reduce the distance between where you make things and where you consume things. There is the bog-standard reshoring, but there is also 'near-shoring' (not bringing things all the way back and instead just putting them slightly less far away), 'friend-shoring' (moving things to a country that may still be far away but is more ally than potential foe) and 'green-shoring' (bringing things back for the primary reason of reducing impact on the environment).

Can we get to zero-emissions logistics?

If we must continue to move materials, components and products over long distances, what can we do to reduce the environmental

* 'Moving a business or part of a business that was based in a different country back to its original country' – https://dictionary.cambridge.org/dictionary/english/reshoring.

impact? Strategic roadmaps and simulations galore have been created to show all the different ways that the pollution generated by land, air and sea transport could be reduced.[23] For ship-borne transport – the primary way in which we shift goods over long distances – these include going back to wind power, converting to electric or hydrogen propulsion and adding cleaning widgets to ships' exhaust pipes. There's also a lot of effort going in to reducing the pollution caused by ships as they sit at dock, and the running of all the associated machines to load and offload the cargo.

Being the nerdy person I am, I have enjoyed reading many of the strategy documents and watching videos of zero-emissions ships. I found it rather surprising to learn, however, that the most significant innovation to be introduced so far to the world of maritime shipping, the change that has had the greatest impact, is remarkably simple.

Tell the ships' captains to drive a bit slower. That's it.

Analysis showed that slowing down by 20 per cent would reduce CO_2 emissions by 34 per cent.[24]* This, of course, means that things would arrive a bit later. Does this matter? We've got very used to the speed at which things arrive, and it was assumed that this was something that it would be hard to persuade consumers to give up, though it actually turns out that many people don't mind waiting a few extra days for things to arrive.[25]

What about the other modes of the global intermodal transport system? Rail remains one of the cleanest forms of transport with approximately 24 gCO_2 per tonne-km compared to 137 gCO_2 for road trucks and 1,036 gCO_2 for air cargo.[26] We are also now seeing the emergence of electric and hydrogen-powered trucks. Both the UK and the US governments have committed to 100 per cent zero-emissions trucks by 2040, with the slight difference between

* And apparently also deliver the additional benefit of nearly 80 per cent fewer 'fatal whale strikes'. That's got to be a good thing.

the two being the US setting up a $7.5 billion infrastructure investment fund to make the plans a reality while the UK has committed a more modest £200 million.[27]

And then there's air freight. While there is a lot of work being done to reduce the climate impact of aviation,[28] this issue presents a complex set of issues. How do we trade off environmental damage from airfreight against a good source of livelihood in poorer communities? This was neatly summed up by Claire Melamed from the charity Action Aid.

> The trade of fruit and vegetables from Africa to the UK accounts for only 0.1 per cent of all the UK's emissions. Therefore, banning organic green beans from Kenya or mange tout from Zambia, say, is not going to make much difference to the UK's overall carbon footprint. However, there are many poor people in Africa who depend on that trade, so, for them, banning organic air freight means fewer children in schools, no investment in small businesses, less development of the economy and more poverty.[29]

Our whole manufacturing logistics system is now under increasing scrutiny, and targets have been set for making all elements of our sea, road and air freight systems net zero over various (and often disappointingly flexible) timescales.

So, now you have your answer for those aliens when they ask you the question posed at the start of this chapter:

We move stuff around so much because we are desperately trying to be very efficient and reduce waste to deliver the things needed to live our lives. This has improved quality of life for millions of people. But in doing so we've inadvertently built a system of making and moving things that is very fragile and damaging to the planet. We keep using this system because we quite like having rapid access to a ludicrously wide variety of products at low cost.

I suspect this might cause the alien to raise their eyebrow (or extraterrestrial equivalent) and consider adding us to the list of planets with populations who have lost the plot. However, I'd like to think that putting planet earth on that list would be premature because, as touched on earlier and as we shall see more in the second half of this book, things *do* seem to be changing for the better.

Before we explore ways this system is now being transformed, we need to look at one more thing. The final part of the manufacturing world we need to understand relates to one basic question: how do manufacturers find out what you and I need or want, and hence what to make?

To start finding the answers to that question, I want to show you something amazing that happened behind the scenes as the UK prepared for the coronation of King Charles III in 2023.

4: Sate

How do manufacturers figure out what we want?

It hadn't happened for seventy years, so I reckoned it was worth watching. The coronation of a new monarch in the UK is, whatever your views on royalty, quite a thing. The UK might be struggling slightly to find its place in a changing world, but we still have it nailed when it comes to pomp and circumstance. Behind the synchronised movement of red, gold and silver, beyond the human- and horse-related logistics of getting the would-be monarch perched on a suitably grand seat at the right time, wearing the appropriate clothes, ready to receive a weighty bejewelled hat, sat some extraordinary manufacturing activities.

How that manufacturing was done illustrates something important about how those who make things interact with their customers.

Sometimes, customers know exactly what they want

Question: Who makes the ceremonial uniforms for the UK military? Answer: One UK manufacturing company called Firmin House, founded in 1655.

Firmin House is now run by members of the Kashket family, whose Russian forebears made felt hats for Tsar Nicholas II. Revolutionary closing of their main market forced them to seek new opportunities, and they moved to the UK. The company now has multiple lines of business that span bespoke tailoring to body armour, medals to buttons, and uniforms to 'horse furniture'.* It is their work on ceremonial uniforms that is most relevant for this chapter.

The making of uniforms for the coronation of Charles III shows one of the simplest ways manufacturers decide what to make for their customer. The customer, the UK Ministry of Defence in this case, called up Firmin House a few months before the coronation and asked something along the lines of *Can you make the ceremonial uniforms for six thousand servicemen and women, ready for 6 May 2023?* Mr Kashket and his team then scratched their collective heads, looked at how many machines and workers they had (their production capacity), talked to their suppliers of red cloth, brass buttons, gold braid, and other rather specialist bits and bobs (their supply chain), reviewed all the other things in their order book (their current customers), discussed prices, and made a decision about whether or not to take on this job. Unsurprisingly, they decided it was worth doing, and the gloriously vibrant results of their hundreds of hours of work were evident to the millions who watched the coronation all around the world.

More often than not, however, manufacturers do not receive such precise instructions. Very often customer demand is uncertain; customers like you and me are unclear about what we actually want. As a result, manufacturers must try to guess what the final customer might want and when they might want it – and that is very hard to do accurately and consistently.

As we know, a factory runs best when everything is smooth, synchronised and predictable. If you know exactly what and how many things you need to make, your suppliers are happy because they know what to deliver to you and when. Predictability allows

* Yes, I had to look this up. It is 'the decorative metal attachment to leather harnesses and saddlery' – https://www.firminhouse.com/products-services/horse-furniture/.

the factory managers to get everything working in harmony and the shop-floor workers to find better ways to complete tasks, so productivity improves. In short, everything gets better. The only fly in the ointment preventing this all working seamlessly is the pesky customer. If only we weren't all such a bunch of irrational, inconsistent and easily influenced people . . .

So how do those who make things work out what customers really want? To start, let's stick with a clothing-related example and head to a small village in pre-industrial revolution England.

From bespoke to in-bulk

Life back then was simpler. Not necessarily better – the average life expectancy was around forty years,[1] spent in squalor and poverty – but simpler. In that context, imagine your trousers were threadbare and you needed new ones. You had a choice of either making new ones yourself, assuming you had suitable skills and access to the required materials and tools, or finding a tailor and then either bartering for or buying new trousers. You either had the self-sufficiency or bespoke tailoring experience. When something got damaged, you repaired rather than binned it.

Then along came the first two of those industrial revolutions and things changed rather dramatically. Thanks to the harnessing of water and steam power and the development of machine tools and interchangeable parts, factories could be built that churned out huge volumes of products. To ensure that there was a market for what the factories were making, it was no good just waiting for people to turn up when they needed something – they had to be persuaded that they needed your product. In simple terms, consumerism and 'planned obsolescence'[2] had to be invented to ensure that all those things being made ended up being bought.

In today's world, for some, the homemade approach is still an option. Others with sufficiently deep pockets can avail themselves of services provided by the likes of Savile Row tailors to have made-to-order, bespoke tailoring, but for the vast majority of customers, whether buyers of *haute couture* or fast fashion, clothes are made and then sold.

And to do that well requires some really sophisticated thinking.

Why is it so hard to balance supply and demand?

In their song 'Keep the Customer Satisfied', Paul Simon and Art Garfunkel very neatly encapsulate what sits in every manufacturing manager's head. However, they were a bit light on why customer satisfaction is so hard to achieve, and did not, sadly, go into any depth on the business strategies needed to deliver it.

So, why exactly is it difficult? To answer that, I want to put you inside the heads of four different people and go through a contemporary shoe-related example.

The customer

You bend down to put on your shoes and realise they not only look a bit scruffy but are actually falling apart. You're a bit sad as they are your favourite; they're shoes that have served you well. With a sigh – realising these are beyond repair* – you accept that you need a new pair. Assuming you decide not to buy online, you want to be able to walk into a conveniently located shoe shop and have your perfect replacement shoes – right style, colour, size, price – available on the spot, handed to you with a smile in a pleasant retail environment.

The shop manager

You want a steady stream of customers coming in and all your staff working at full capacity serving them. You don't want to be paying

* A decision typically reached much more quickly today than in the past. We'll come back to this issue in Chapter 8.

rent for space to store a large quantity and variety of shoes if there is a risk that they might not sell quickly. The shop's equipment – seats for trying on shoes, foot-measuring doohickey most customers haven't used since childhood, point-of-sale equipment – would be in constant use and taking up minimum space. The stocks of shoes and shoe-cleaning products would spend almost zero time in the storeroom or customer-facing shelves before being purchased. However varied customer demand might be, it would ideally be matched perfectly every time by what has just been delivered by your supplier.

The shoe supplier

You want your factory operating smoothly within minimal variation.* As most of us do not live in socialist, state-planned economies, where perhaps there could be one approved shoe design and the only variation would be size, your shoe factory must produce a wide range of different styles in anticipation of particular customer preferences. What you really want is to produce a run of a particular style of shoe and know that this whole batch will be shipped and sold via some large, easily accessible retail facilities with significant capacity for storing stock. You don't want to be on stand-by for the instant making and delivery of a very specific style, colour or size of shoe to address specific customers' unique demands in thousands of different retail sites scattered all over the place, and you absolutely don't want shops sending back shoes that didn't sell.

The shop worker

You want a nice, steady, calm workflow, with everything needed to satisfy the customer needs immediately available. You might appreciate time to chat with customers and colleagues, with pauses in the flow of customers to allow you to sort out various admin

* We are assuming your factory is not one that is hand-crafting bespoke shoes to order.

tasks. You don't want to deal with frustrated customers who have to be told, *Sorry, we don't have your size in stock. I think we can order a pair for you but it'll take a few weeks.*

You can see what the problem is. These activities are pulling against each other. That simple example shows the tensions that sit at the heart of any manufacturing business: how to balance supply and demand across factory operations and supply chains and between makers and consumers. Having previously focused on the *supply* side of this tricky balancing act, let's dive further into explaining why this *demand* side is so frustratingly hard to get a grip on.

In trying to understand what customers really want, every manufacturing firm is trying to solve three related issues:

How many things should I make?

Which things should I make?

How might this change over time?

Everything we have discussed on designing the production process and working out where things need to be made can only happen if these three issues – volume, variety and variability – are understood. To do that, manufacturers have to guess what will be going on inside customers' minds at some point in the future.

Welcome to the pseudoscientific world of demand forecasting.

Forecasts are always wrong

I used to review student business plans submitted to an annual entrepreneurship competition. Engineering students were particularly good at presenting detailed spreadsheets and beautiful graphs of the scale of demand for their as-yet-unbuilt products. These were clear, elegant works of pure fiction. Why? Because there was no *real* data on how many customers would open their wallets and hand over cash for this amazing new gizmo that goes 'beep'. They might have asked a sample of would-be customers, *If this existed, would you buy it?*, but data gathered from questions like that is notoriously unreliable, especially for never-before-seen innovations. The only truly reliable data comes from a customer who has *already* handed over money in return for a product or service. Until then, everything is basically guesswork – informed, sophisticated and data-driven guesswork, but guesswork nonetheless.

So how do manufacturers come up with predictions of future demand, and how do they get to a position of sufficient confidence to start making a particular product to meet that demand? They use a combination two types of crystal ball.

Learning from history

The first crystal ball relies on real, measurable data based on things that have already happened. Cadbury, the confectionary company, knows exactly how many of each of its products it has sold over many years and can make pretty strong assumptions based on these numbers. For instance, they could see an upward trend as they sold 9.72 million Creme Eggs in 2019 compared to 10.86 million in 2021, followed by a slight drop to 10.69 million in 2023[3] – this movement may have been influenced by the pandemic, with people eating more chocolate during lockdowns. Similarly, Airbus keeps precise data on trends in their order book.[4] You can see the clear peaks and troughs reflecting the ever changing patterns of demand for different sizes of aircraft in response to a complex range of social, economic and geopolitical factors – these could include political attitudes to the free flow of people, actual or predicted fluctuations in fuel prices, and changing consumer attitudes to the sustainability of different forms of transport.

Such historical data can then be used to make predictions about what customers might want in the future. These vary from the extremely crude – such as expecting customers to do the same as they have done in the past, appropriately known in marketing circles as the *naive approach* – to the quite sophisticated. The more advanced approaches use some mathematical manipulations that try to predict the next value from a series of numbers.[5] While past performance is not a reliable indicator of future results (just as financial-investment advertisements warn), these quantitative methods are still useful for manufacturers working in reasonably predictable markets. They allow them to make pretty good estimates about how much of a product might be wanted and can help ensure that their factories and suppliers are all geared up to respond to this demand.

But what do you do when there is only limited past data on which to draw? This is what faces any manufacturer who plans to introduce a new product.

Guessing the future

For some new products, manufacturers can say, *Well, this new X is a bit like the old Y, so we'll use the data for Y*. As I typed this paragraph in June 2023, Apple had just announced the 'Vision Pro', their virtual reality (VR) headset. How do they forecast how many customers are likely to want such a product? There are other VR headsets on the market, so they have that data on which they can draw,[6] but Apple's product is different. How do they know whether past sales data relating to those other products provides a reliable indicator for sales of their fancy new gadget?

The truth is, they don't.

Sometimes manufacturers are introducing a product that is completely new, something that has never been seen by customers before. That makes getting confidence in predicting what customers want really difficult, and something that might be stuck on your fridge or the edge of your computer screen illustrates this quite clearly.

In 1968, a scientist called Spencer Silver was working at the US firm 3M, trying (and failing) to develop a super-strong glue. He ended up creating something that was merely tacky rather than really sticky. A few years later, one of his colleagues, Art Fry, found a use for this low-grade glue: it could temporarily hold bookmarks in his hymn book. Thinking they were onto a winner, the two of them joined forces, had batches of these 'bookmarks' made and gave them to their colleagues . . . who were not impressed. *Why would anyone pay money for sticky scraps of paper?* After they had given out more free samples, however, people gradually realised how useful they were and, seemingly out of nowhere, the Post-it became a very commercially successful manufactured product.[7]

Both the teams developing Apple's Vision Pro and 3M's Post-it will have tried to build up an understanding of the potential market for their new products in the same way: asking people. There are two options for doing this: ask the people you think might buy your product, or ask smart people what they think buyers might want.

The first of those two options sounds great. What could possibly go wrong with just asking potential customers? Well, type 'failed product launches' into Google and you'll get a pretty clear answer to that question.*

The problem is that most of us aren't that good at imagining what we might want in the future. Whereas we might be pretty good at responding to questions like *How many pairs of jeans might you buy over the next twelve months?*, we really don't have a clue when it comes to completely new things. A quote attributed to

* I've put a few classic examples on the book's website, www.timminshall.com.

Henry Ford sums it up: 'If I had asked my customers what they wanted, they would have said a faster horse.'[8]

For the second option – asking smart people what they think buyers might want – there is the task of actually finding these smart people. Manufacturers have several ways of doing this. They could ask their own sales team, the people who probably know the most about their customers, what new things they think customers might want. Or they could do what is called *scenario planning*, where they create an imaginary situation and ask, *If this were true, do we think people might want our product?* For example: if oil prices were to suddenly spike by 50 per cent, might this increase demand for electric vehicles?

And then there are so-called 'Delphi studies'. Rather than trudging all the way to Greece and climbing Mount Parnassus to find a mythical priestess to consult, a Delphi study is a structured process for consulting a wide pool of experts. These different views are then averaged out over rounds of questioning until you end up with a reasonably well-informed consensus view of the future.

So, there are two different types of unknowns relating to customer demand that manufacturing firms have to deal with. Manufacturers may be wanting to understand a market for a new product, one that potential customers may never have seen before. Then there is trying to understand how the market for an existing product might change in response to all sorts of factors.

If we assume that people who run manufacturing operations – people like Alison at Fitzbillies or Ian at Palm Paper – can get a sense of what future demand for a product might be, what happens if it is too big or too different for them to deal with? What if it looks like it might be a temporary spike in demand, responding to which might cause all sorts of short-term problems for the factory? What if all this analysis predicts a temporary lull in demand that means people and machines will be standing idle?

The likes of Alison and Ian are not passive players in this game. They have some tricks up their sleeves.

How to manage demand

There's a particularly awkward type of silence that engulfs a large lecture theatre when no one is responding to a question you've put to the audience. It's so tempting to leap in with the answer. Silence stretches the seconds. You feel your heartbeat in your ears.

'Soup?' says an uncertain voice from somewhere in the void, mercifully easing the collective tension.

That voice was giving a great answer to the question, 'What could a food manufacturer that makes things that sell well in the summer – like ice cream or prepared salads – make and sell during the winter?' The answer given is what's known as a *counter-cyclical product*. In this example, the cycle is dictated by the seasons, but a cycle could also be dictated by the peaks and troughs of an economy or the timing of significant events (like Christmas, a major challenge for toy manufacturing companies in particular).

Making counter-cyclical products is one example of the many ways in which manufacturers can move beyond just being passive responders to what they think customers are demanding and try to alter our behaviours: they can move to what's called *demand management*.

The simplest and most clearly visible form of demand management is advertising. People don't seem to be demanding your product? Run an advertising campaign or get social media influencers to increase awareness and hence demand.[9]

Pricing and reserving are other ways of manipulating customers. Is demand forecast to surge? Manufacturers can bring in peak-time 'premium pricing' to encourage some customers to delay their purchases, smoothing fluctuations in demand to allow the factories to maintain a reasonably steady flow.

But even with tricks like these to try to control or manipulate demand, manufacturers still have to make sure that they are able to respond to changes; they need to know how all those things we talked about previously, within the factory and along the supply chain, are able to be ramped up or dialled down.

This is the world of *capacity planning*.

How to manage supply

The textbook authors Michel Baudin and Torbjørn Netland pithily sum up what faces most manufacturing firms seeking to understand customer needs: 'except in rare, special circumstances, we never exactly know what the demand will be in the future. There is no general answer that pops out of a mathematical formula.'[10]

One of those 'rare, special circumstances' would be the coronation of King Charles III as described earlier – Russell Kashket got an unambiguous signal from his customer. Yet even then, Russell and his team will have had to do some pretty swift calculations to work out whether these clear requirements could be met. On one hand, they could do a quick back-of-an-envelope calculation, drawing on their knowledge gleaned from years of experience of running the operations of Firmin House and building up relationships with their material suppliers. On the other, they could take a more systematic approach and try to do some modelling or scenario planning based on all the data available, drawing upon their understanding of the steps required to make their current products and the known capacity of their suppliers to deliver constituent parts.

In practice, most manufacturers will manage this linking of potential sales with what the firm can actually deliver using a combination of these approaches. This has a name: *sales and operations planning*. Spreadsheets will be populated and pored over, graphs and bar charts will be squinted at. In successful firms, this won't be left in the hands of a separate team of spreadsheet squinters but will (or rather should) be done by people from all parts of the business, from sales, HR and finance to those actually managing the production, drawing on a wide base of knowledge.

Even with this holistic perspective, though, sales and operations planning activities often do not deliver a single clear view of what factory managers actually need to do. Are the forecasts telling the managers that more of an existing product is needed? Or does a slightly modified version need to be developed, or even a completely new product? Is this a one-off, random surge in demand that will pass, or is this an ongoing upward trend?

How do companies use that data to make decisions about what to actually make?

Deciding what and how many to make

Sitting on top of the sheets of paper scattered before me is a Uni-ball Eye Micro UB-150 pen manufactured by the Mitsubishi Pencil Company Ltd of Japan.* I have a suspicion that this will have been made using an approach called *level production*, which assumes that, even if demand goes up and down throughout the year, production can be kept at roughly the same level. In months where sales drop, the Mitsubishi Pencil Company can go on making pens and add them to their stock. Then, in months when customer demand exceeds the standard level of production, they can use their stockpiles of ready-made pens to respond to this higher demand without having to suddenly ramp up production. This works well for things like pens and dried pasta (and some bicycles) which have quite a long shelf life, but

* Those sheets are covered in scribbles about things to add to the book that, due to my terrible handwriting, were not included. If you think you know what I meant by: 'choice + make uv not wedlybink > fasting yam', let me know and I'll be sure to add it to any future editions.

less well for more perishable goods like fresh cream cakes and sushi, and even with non-perishable goods there is always the risk that customer demand changes, leaving you with a load of things you can't shift.

An alternative approach to level production is *demand chasing*. This is a little more nerve-racking for the production planner as it relies on high-quality forecasting and having machines, people and suppliers respond at gazelle-like speed to changes in demand. The basic idea is that you look closely at a graph you've plotted of anticipated patterns of demand and then try to ensure that the output of your factory tracks as closely as possible to every bump and dip in that line. This avoids you needing to spend money on storing finished but unsold products.

Perhaps unsurprisingly, demand chasing is really hard to do well. The target is always moving, and the information provided from your crystal balls is never going to be 100 per cent accurate. Demand may be higher or lower than you predicted, or come sooner or later than anticipated. Take the example of the aircraft industry where things were particularly 'exciting' in the immediate post-pandemic period. In 2023, there had been a 440 per cent year-on-year increase in aircraft orders. The backlog of aircraft orders reached a record high of 14,535, representing more than ten years of manufacturing work and an estimated value of £219 billion to the UK alone.[11] While it might be great for the likes of Boeing and Airbus to have such healthy order books, they have to be able to actually make and deliver all those planes – without compromising on quality.

Manufacturers need to be confident that their production capacity – the ability to make things based on the people, machines and materials they have access to – can cope with such violent and rapid changes in demand. You could try to rapidly expand by building new factories, buying new equipment, adding more suppliers and recruiting more staff, but having done everything needed to respond to a surge in demand, what happens when demand slows down and you have too much production capacity? It'll be tough to keep paying for people, machines and materials that are now not needed, but is it worth keeping them in anticipation of the next surge in demand? If you reduce your production capacity now, are you confident that you'll be able to ramp things up again in the future? Are there other things that these people and machines could be used for?

These are the tough decisions that need to be made in the world of sales and operations planning.

So, there we have it. In order to understand what to make, manufacturers try to not only forecast demand but also nudge consumers towards certain choices. Tightly coupled to all this 'demand side' work are the 'supply side' activities – how to ensure that, whatever the demand is, the manufacturer is able to keep those customers satisfied.

Unfortunately, as is the way in the modern world, things have got a bit more complicated. One of the biggest issues is that many manufacturers now have access to too much data.

How digitalising data made things better, and worse

Up until the 1970s, computers were big centralised machines that were delicately cared for by highly skilled technicians in air-conditioned rooms somewhere far away. After some clever people in California managed to fit a whole computing system onto one tiny bit of silicon – the microprocessor – it triggered the creation and explosive growth of the personal computer (PC) market.

The PC democratised access to data processing so that even the smallest company could have access to computing power to process data and potentially improve business. However, you can only process data if you can access data and, as we know, getting useful data on what customers want is notoriously difficult – especially before all those computers were joined together to form the internet.

Before the arrival of the internet, the most important places where you could get information on what customers were doing was at the cash register – the actual point of sale. Jump forward from the 1970s to the 1990s and retailers were able to use the democratised computing power of the internet-connected PC to access more customer data directly and reliably through the use of 'electronic point of sale' (EPOS) technologies.[12] This was a significant step forward in allowing manufacturers to better match supply and demand and helping them to pull off that clever trick of demand chasing.[13]

This, however, created a new problem: there is now *a lot* of data. Approximately 328.77 million terabytes (or 0.33 zettabytes) of data are created *each day*; over the span of twenty-four hours, 333.22 billion emails are sent; and at the time of writing, it was estimated that 90 per cent of the world's data was generated in the previous two years alone.[14] This creates a problem called the 'signal to noise ratio' – the gems of useful data you want are now buried deep within those zettabytes of mostly irrelevant data.*

For manufacturing firms, the availability of this digitalised data also presents an amazing opportunity. We talked earlier about how firms need to draw on past data to work out what things customers are likely to be demanding. In the days before everything was networked and integrated, someone would have been tasked with providing monthly or quarterly sales data, laboriously captured and manually typed into spreadsheets and circulated to those working

* And there is the related problem of the *quality* of the data. With such vast quantities now available, how can you check whether every data point is actually showing what you think it is showing?

in sales and operations planning. Now, there is a near-constant stream of data available.

The generation of this massive amount of data has been made possible thanks to the digitalisation of almost every aspect of our lives. We are now able to get a much higher level of detail of customer trends – and at a much higher speed – than in the past. You, me and every other consumer connected to the internet in some way is benefiting from this extremely sophisticated approach to connecting what we want with how it will be made. And, whether we know it or not, we are all a key source of input to this amazing system. Me scrolling through the Amazon app on my phone to order some more of those pens from the Mitsubishi Pencil Company – or any other product for that matter – is providing one teeny-tiny extra bit of information on my needs, wants and preferences to help manufacturers develop a sort of creepily accurate 'digital twin' of myself.

The scale and richness of this data has provided manufacturers with the ability to make better decisions about what is wanted and how they can best respond, but the phenomenal increases in performance and reductions in the cost of computer power have also enabled *faster* responses. It has allowed demand chasing to be more speedily and effectively managed.

The fact that manufacturers can now do this is great, but it has also led to consumers expecting this as the norm, and the more we come to expect that whatever we want will be available all the time, the more of a shock it is when things go wrong.

How to respond to ever-more-demanding customers

In parallel with improvements in how manufacturers understand what we want, they have come up with some rather clever ways of ensuring they can respond rapidly to our irritatingly specific yet varied demands. These can be summed up as what, who and how.

What

One thing that manufacturers can do is to design products that are easy to adapt to different customer requirements. To illustrate this, I'd like to start with some old jokes about one brand of car that will only make sense to people who grew up in the UK in the 1970s:

> Question: What do you call a convertible Skoda?
> Answer: A skip.
>
> Question: How do you double the value of a Skoda?
> Answer: Fill up the petrol tank.

Many of you might be puzzled by these 'jokes'. Skoda make excellent cars. But back in the 1970s, when the Iron Curtain was still firmly drawn across Europe, one of the few products from the communist side of things that those of us living in Western Europe got to see were Skoda cars. While they were simple, reliable and cheap, they did lack a certain lustre on the branding side of things and they did not hold their value well. Fast forward through the 1980s, 1990s and collapse of the Eastern Bloc, and Skoda became part of the mighty Volkswagen Group. The engineers at Volkswagen then did something rather clever: to help them respond to the widely diverse demands of car buyers, they decided to use a *platform strategy*.

A platform strategy for car design describes an approach used by manufacturers to develop and produce multiple different vehicle models based on a shared set of components and technologies. This is a very cost-effective way of producing a range of cars that look different to the customer but whose major structural components – the basic architecture – are common to all. And what's particularly relevant for the theme of this chapter is that this strategy allows car manufacturing firms to respond to a broad range of different customer preferences. There's no need to spend vast amounts of money to build a completely new car from scratch as customer preferences change; manufacturers can instead respond to changes in customer demands by swapping components and features in and out, without needing to go through the expensive process of changing the whole underlying product design. Volkswagen Group's *Modularer Querbaukasten* (MQB), which underpins various models across the VW, Audi, SEAT and Skoda brands, is a notable example of successful platform strategies in the automotive industry, as is the Toyota New Global Architecture (TNGA) used by Toyota and Lexus, enabling them to share components and technologies while maintaining 'brand-specific' features that customers in different market segments demand.

So, the next time you are sitting in traffic and are bored, take a moment to look at any passing VWs, SEATs, Audis and Skodas. You'll now notice that, if you squint a bit, across each car badged with those different brands, there are remarkable similarities. That is what allows Volkswagen, as one of the largest car companies in the world, to respond cost-effectively to ever-shifting and complex patterns of demand from their customers.

Who

Manufacturers can also use creative ways to organise people, machines, materials and processes to respond to fluctuations in what customers want. One tried and tested way is to temporarily get someone else to make things on your behalf, to cope with what might be a short-lived change in demand. In fact, before the First Industrial Revolution and the idea of the integrated factory – everything being brought together to make things in one place – production was commonly distributed among lots of individuals in the local community. This form of decentralised production was referred to as the 'cottage industry' or the 'putting-out' system.

Under the putting-out system, a merchant would provide raw materials and instructions to rural households where members of the family would complete some simple production process. The finished goods would be returned to the merchant in exchange for payment. This was very common in the textile industry, where merchants would ship out raw materials such as wool or cotton to rural weavers.

Drawing on the services of other firms to provide some additional capacity when needed is not uncommon today. If there is a sudden surge in customer demand and your current people and machines are already working flat-out, it makes sense to see if someone else with the appropriate skills has some spare capacity that could help you out. During the COVID-19 pandemic, this proved to be a (literally) life-saving strategy. As we'll see in more detail later, when demand for personal protective equipment, hand sanitisers and medical ventilators surged far beyond the capacity of the main manufacturers, an extraordinary thing happened: manufacturers from non-medical sectors found that they could repurpose their operations to help the medical firms make more of the required products. That traumatic but ultimately successful experience delivered valuable lessons for manufacturing firms. It illustrated the value of learning how to use the resources of other firms to add some 'elasticity' to their ability to respond to sudden changes in customer demand.

How

There is a third way in which manufacturing firms can prepare themselves to be able to respond to fast-changing demands. Rather than having machines optimised for making one thing really well, why not use machines that are much better at making lots of different things? That way, your factory will be able to respond to whatever maddeningly inconsistent requirements your customers send your way. And there are some technologies that can help you do that.

There are broadly three kinds of tools and machines. First, there are those that are designed to do just one very specific thing. At the extreme end of specificness, Firmin House use a very special set of tools just to make the shiny helmets for cavalrymen and cavalrywomen.*

Then, there are tools that do one type of thing for many different applications. A drill, for example, is very good for making straight, round holes, and depending on the size and material of the drill bit you can make all sorts of holes for all sorts of purposes – but these holes will be round and straight.†

The final type of tool is something that can be made to make almost any shape, and one example is a 3D printer. These machines offer designers very high levels of what is called 'design freedom'. If you want a component that has a straight round hole going through its middle, a drill will do the job fine, but if you want that component to have an oval hole that starts at one side and goes round a corner to become square, a drill is useless. A 3D printer can print an object with such a hole running through it. And if the designer wants a hole that starts as a hexagon and ends up as a triangle, that's no problem – the high level of design freedom offered by a 3D printer means that you just change the design on the computer and send a new print file to the printer.

That might sound quite impressive, but what has that got to do with responding to changing customer demand? Well, imagine you are the manager of a factory that makes components for one

* Rather incredibly, one of the tools – a blacksmith's elm – used for shaping the main part of the helmets worn by members of the Household Cavalry has been in continuous use since the formation of Firmin House in 1655 (page 105 in Grant, P. (2024). *Less*. William Collins). The blacksmith's elm works, and they only need one, so they have kept using the same very specialist tool for that single task.

† OK, before anyone starts shouting at me, I know that you can actually drill a square hole using a mortiser drill, but *most* drill bits are for making round holes.

of the big aircraft companies who are trying to respond to that 440 per cent year-on-year increase in aircraft orders we mentioned earlier. You have your whole factory optimised for making very specific components, with specialist machines bought and set up to make very precise parts. How can you respond to this surge in orders? Your machines and workers might already be operating at full capacity. Do you rush out and buy more specialist machines so that you can cope with this order? And what happens after the order is completed?

As we know, you could end up with a bunch of expensive machines and specialist workers sitting idle as demand returns to pre-surge levels. However, what if you had some machines that were much more flexible, that could be temporarily set up to make certain parts but which could also be used to make something else that might be demanded in the future? That's where 3D printing technology can – in some specific cases – be really useful.

But manufacturers can also do something much more radical. How about not selling a product at all?

How to deal with the fact that 'no one wants a drill. What they want is the hole'[15]

This quote relates to something that has now been given the linguistically inelegant name 'servitisation'. It's a very simple premise, but one that has transformed the way many manufacturing (and many non-manufacturing) businesses operate: you don't focus on selling the customer your product, you focus on addressing the customer's real needs, on solving their problems.

This is best explained with an example.

An airline doesn't really want to buy and be responsible for maintaining jet engines for its aircraft. What it really wants is to have aircraft that are always available to whisk their passengers off to wherever they want to be. With the servitisation approach, jet engine manufacturers like Rolls-Royce or GE offer 'power by the hour', where the airline pays the manufacturers to power a certain number of hours of flight per year. The airline doesn't have all the faff of maintaining and repairing the engines – that's all sorted by the jet engine companies. This can be good for everyone. Why? Well, in the past, jet engine companies made a substantial amount of money from 'spares and repairs' for engines they sold to airlines. With the servitisation business model, the jet engine companies now have a huge incentive to provide low-maintenance, reliable engines, as they themselves will have to pay for any repairs or maintenance required. On the other side, airlines don't have to worry about managing workshops and training technicians – all that is taken care of by the jet engine company. The airline can focus instead on investing in the things they are good at: trying to make the experience of sitting in a pressured aluminium tube travelling at nine hundred kilometres per hour at ten thousand metres in the sky quicker, cheaper or just a bit more pleasant.

This happens in other sectors, too. The car industry has long had the idea of allowing customers to lease rather than buy cars. This has more recently been extended into the idea of car clubs – like Zipcar or Getaround – and other models in which you just have access to a car when you need it, rather than having one expensive tonne of metal, plastic and rubber sitting idle for the vast majority of the time.*

But whether we are talking about aircraft or cars or any other products, there is one essential resource without which servitisation cannot work: data. Rolls-Royce and GE need exquisitely detailed data on not only what the customers are doing with their engines, but also what the engines themselves are up to; Zipcar and Getaround need data on where their cars are, what condition they are in and which customers are using which cars. How this

* In the UK, cars and vans spend 96 per cent of their time parked. https://www.racfoundation.org/wp-content/uploads/standing-still-Nagler-June-2021.pdf.

is achieved is truly remarkable, and this is something we'll return to later in this book.

You are now equipped with the basic tools and language to understand how things work within the manufacturing world today – and why they are the way they are. But the way we manufacture things is in a process of constant change. Next, I want to show you how those who work in the world of manufacturing are not only dealing with daily battles of production and logistics but managing a whole set of seismic changes that will affect the way we make, move and consume things. To understand how that future is transforming the present, we need to understand how change, how *innovation* happens in manufacturing.

And to do that, the first step is to go and make a cup of tea.

PART TWO

How the World of Manufacturing Is Transforming

5: Change

How Manufacturing Can Transform the World – and Vice Versa

Click. That's a specific sound you probably hear several times a day. It is heard hundreds of millions of times a day around the world. It is the sound of a flow of electrons automatically being halted at a precisely defined moment to ensure the safe and sustainable operation of a machine which is vital for our well-being. It is the click that announces that the water in your kettle has boiled.

That sound is generated by a small, shiny, circular device, just under two centimetres in diameter, buried within almost all the world's electric kettles. I have several of them sitting here on the kitchen table beside my laptop, some in their raw state, fresh from the factory, and others embedded within the plastic assembly that makes up the control system for your kettle.

The upper and lower surfaces of these specially shaped discs are made of different metals, each of which expands at a different rate when heated. As a result, when the heat being carried by steam in your kettle passes over this small disc, one side expands faster than the other, until it flexes with a little 'pop'. That popping movement is strong enough to act as a switch to halt the transmission of electricity to your kettle's heating element.

That is what happens every time you – and over a billion other people all around the world – use an electric kettle to boil water.

How kettles learned to switch themselves off

I have a sneaking suspicion that you, like me, had not given much thought to how your kettle switches off. You might also not have spent much time thinking about how we ended up with every electric kettle in the world knowing when to turn itself off. The story of the development of this tiny device in fact gives a great illustration of how change happens in the manufacturing world.

To tell that story, I need to take you back to the 1970s where a quiet revolution was about to take place in the world of kettles. This might not have had quite the same impact on headlines as other issues of the time,* but nonetheless it was going to have an impact on billions of us.

Up until the 1970s, kettles were made of metal, and came in two varieties. One used an approach largely unchanged for millennia: you filled the kettle with water and applied heat to its underside. The heat source may have moved from being an open fire to a gas or electric hob, but little else changed. Then, in the late nineteenth century, the first standalone electric kettles started appearing on the market. The early models had heating elements separated from the water as engineers hadn't quite worked out how to avoid the exciting flash-bang outcomes when mixing electricity and water in a domestic appliance. We had to wait until the 1920s for technology to develop to a point where the electric heating element could be safely immersed in water.[1] From then, the design of electric kettles pretty much stabilised. You added water to the kettle, plugged it in and switched it on, and when you heard a continuous rumbling and saw steam jetting out of the spout or heard the associated whistle, you knew your water had boiled. If you got distracted and wandered off, the metal kettle would simply continue to boil until all the water had evaporated, and the heating element would then burn itself out. Not good, and it would leave an acrid smell in the air, but it probably wouldn't be life-threatening.

* The Cold War, Roe vs. Wade, first female Prime Minister in the UK, breakup of The Beatles, to name a few.

However, if your kettle was made of plastic, letting it boil dry would likely result in a bit more drama. The heat generated would eventually melt the plastic body of the kettle, causing it to collapse onto the heating element, resulting in clouds of toxic smoke, flames, and probably your kitchen, office or hotel room ending up as a burnt-out shell. As a result, despite the advantages of a plastic-bodied kettle,* manufacturers were reluctant to make and sell them.

To make plastic electric kettles safe, some kind of switch was needed that would automatically turn the kettle off once the water had boiled. And to deal with people like me who struggle to engage with the complexities of life first thing in the morning – people who might skip the important 'add water to kettle' step before turning it on – another switch was needed. This second one was needed to make sure things shut down automatically long before the melting point of the kettle's plastic body was reached.

The bimetallic discs described at the start of this chapter are the core component of a remarkably elegant solution to these problems. And this is where a man called Dr John C. Taylor, the inventor of those little shiny devices, enters the story.

The machines that make switches by the billion

It's early on a grey spring morning in 2023 and I'm at London City Airport, looking out at the tiny Loganair plane which will be taking me on the short hop to the Isle of Man. On board, peering over tartan headrests, I guess my fellow travellers to be a mix of tourists, returning residents and, as I find out later, investors from mainland China. The propeller engine sitting disconcertingly close to the left side of my head clunks, whirrs, and whines into action. Both engines rev up and we bump along the taxiway before turning to lift off over the waking city of London.

* Lighter weight, complete mouldability and lower cost of production.

A cup of coffee and a biscuit later, we descend through clouds, glide over the Irish Sea, turn along the green and brown coastline to gently bump down onto this British Crown Dependency. Walking out through the terminal building, I experience one joy of a small provincial airport: there is only one car waiting immediately in front of the exit. Sitting in the driving seat is Dr Taylor himself.

After lunch at his house, we head north across the island to the town of Ramsey and turn into an industrial estate and park outside a brown, two-tone, low-slung industrial building. This is the heart of the global manufacturing operation of the company founded by John whose main product ensures you can boil the water for your tea or coffee safely and energy-efficiently.[2]

As soon as I enter the architecturally modest manufacturing facilities of Strix, I am told in clear terms that I won't be allowed to take any photos or share much information on what I see. And for good reason. Within this building sit the machines – designed and built by John and his team – that since the 1980s have been mechanically ker-chunking away, producing billions of those perfect bimetallic discs – known as 'blades' – which are now a standard feature of the world's electric kettles.

What I am able to share are the basics of what happens in the Strix factory, which by now will be very familiar to you – like any factory, the Strix one takes some inputs and does some processing (people follow some methods using machines and materials) to convert them into more valuable outputs. The inputs are strips of bimetal coiled around a wooden spool; the outputs are bags of perfectly made bimetallic blades ready to be shipped off to China.

The process starts with the bimetal coils being unspooled into exquisitely engineered, purpose-built stamping and pressing machines. After some heat treating, labelling, testing and packing, they head out the door to be shipped off to another Strix factory ten thousand kilometres away in Guangzhou, China. Why do they have a factory there? Because that's where most of Strix's customers – the kettle firms – make their products. On arrival, these blades will be inserted into the plastic injection moulded switch and power connection assembly buried at the base of a kettle. The assembly is then added to whatever fancy kettle shape market research shows the customers are predicted to want. The finished kettles are then packaged and shipped off to retailers, before ending up in your kitchen, office or hotel room.

For us to have better products, someone needs to invent, experiment with and refine a new way of doing things. In the case of the electric kettle control, this was Dr Taylor and his team designing a complete water-boiling control system – based around a very precisely designed and engineered bimetallic blade. But of equal importance is the design of the process that can make the products in the required quantities, and at the appropriate quality and cost. That process – like those we have explored throughout this book – needs the appropriate combination and synchronisation of people, machines and materials, both within the factory and across the supply network. And all of that making and moving of components and products would be pointless without an understanding of what the customer (the kettle companies) really wants, and what benefit they and their customers (you and I) will get from using the new product.*

But what happens when the new product is going to replace something that is already being made and sold, for which there

* There are also externalities to consider. Strix estimate that five million tonnes of CO_2 are saved each year through the use of kettle control products – equivalent to emissions from one million petrol cars. Their data predicts that two seconds in steam switch-off time saves forty-four gigawatt hours or £9 million annually in the UK.

are already hundreds of firms doing the final assembling of the thirty-thousand-plus components typically required for each product, as well as thousands of suppliers making all those components, millions of staff working for all those companies and tens of millions of customers?

This is exactly what is happening now with the manufacturing of cars as we switch from internal combustion engines (ICE) to electric vehicles (EVs).

This is manufacturing innovation on a planetary scale.

Transforming the world from ICE to EV

I have such powerful positive memories of the Volkswagen campervan we owned during my early childhood that since I became a driver, rather irrationally, I have only ever bought Volkswagen cars. On a visit to the local dealership in 2022, I was surprised to see how much things had changed since my last visit five years earlier. For one thing, an electric car was taking up most of the entrance.

'Ah,' said the salesman as I turned sideways to squeeze into the showroom, 'I see you are interested in the all-electric ID.3 – can I tell you more about this car?'

Here's something I found surprising: at the end of the nineteenth century, as the car industry was in its earliest stage, the bestselling vehicle in the US was not powered by an internal combustion engine fuelled by petrol. It was an electric battery-powered car.*

In the early days of the automotive industry, the relative benefits of electricity over internal combustion to power cars were substantial. Being mechanically simple, electric vehicles were reliable

* The 'Columbia' produced by the Pope Manufacturing Company.

and easy to maintain. They were also smoother to drive, easier to operate and didn't emit clouds of pollutants. The early models used lead-acid batteries, which were heavy and slow to recharge, but for many urban applications they were just fine.

So why did we ditch them in favour of petrol-powered cars, and why have we stuck with these more polluting and complex machines for over a century? According to Tom Standage in *A Brief History of Motion*, it's the result of a combination of technological and psychological reasons.[3]

Petrol has a major natural technological advantage. It has a higher *energy density* – it stores more energy per unit of mass – than a battery. Back in 1897, would-be buyers also expressed the same psychological concerns that many of us have today: range anxiety. If I buy one of these electric cars, can I be sure that the battery will power the whole of my journey, and, if not, will it be easy to find a place to recharge it?

At various points over the past hundred years or so, interest in EVs was temporarily renewed. This was often in reaction to a spike in petrol prices caused by some geopolitical imbroglio,[4] and the pattern of events was fairly consistent. The supply of oil became constrained (or people feared it would be), prices surged, queues formed at petrol stations and demand for smaller, more fuel-efficient cars increased. Engineers thought, *Hey, why don't we develop a car that doesn't need petrol? We should make electric cars!* Rather quirky EVs would be launched, appear in the news and, despite their poor performance and high cost, sales would grow. Then the international incident would be resolved, the price of fuel would drop, people reverted to petrol vehicles and either the EV companies went bust or the large car firms making the EVs quietly withdrew them from the market. This cycle repeated every ten to fifteen years or so.

But then, three things happened.

The first was the development of improved ways of storing electrical energy. A new technology that was lighter, could store more energy and was quicker to recharge than lead acid was developed: the lithium-ion battery.[5] The second was our tragically late realisation that curbing CO_2 emissions to ensure our survival on this planet would probably require us to stop driving around in fossil-fuel-burning mobile power stations. And, luckily for the planet, the rather important third was triggered by two engineers being very unhappy with a decision made by a US car company in the early 2000s.

In 2002, the mighty GM corporation halted sales of its first EV, the not-very-imaginatively named EV1.* This decision to shut down one of the first serious attempts to mass produce an electrically powered vehicle designed from scratch made Martin Eberhard and Marc Tarpenning very cross – so cross, in fact, that they decided to form a company to make and sell their own EV. They decided to name the company after Serbian-American inventor and electrical engineer, Nikolai Tesla.

In 2004, one Elon Musk made a substantial investment in that business and helped push its rapid growth. The firm launched a series of cars, starting with the Roadster, followed by the Model S, Model X, Model 3 and Model Y.

Tesla Motors become a phenomenal success story despite one or two, er, bumps in the road. In 2010 it became the first US car company to sell shares to the public since the Ford Motor Company floated in 1956; eleven years later, Tesla Motors became one of the very few companies ever to be valued at over $1 trillion,[6] and although its value has dropped since then, in the summer of 2024 it still had a higher market capitalisation than the value of the next seven car companies *combined*.[7]

Not bad for a company that only sold its first car in 2008.

* A decision so controversial that it became the subject of a documentary movie – *Who Killed the Electric Car?*.

How 'creative destruction' drives change

Tesla had done what the Austrian economist Joseph Schumpeter hyperbolically dubbed creating a 'gale of creative destruction'. He described this as the 'industrial mutation that continuously revolutionises the economic structure from within, incessantly destroying the old one, incessantly creating a new one'.[8] Schumpeter pointed out that it usually won't be existing firms at the heart of such change: it will be entrepreneurs setting up new businesses, doing things differently, that will make this happen. The founders of Tesla played a pivotal role in whipping up the particular storm that is transforming the global automotive manufacturing industry and moving it away from internal combustion engines.[9]

As a result of this Tesla-induced gale and the starkly clear commercial opportunity now sitting in plain sight, all the world's major car manufacturing companies saw that customer preferences were changing and that they would need to move sharpish to launch their own battery-based EVs.

This begs a question: if the EV market was such a huge opportunity and these massive, well-resourced companies knew the car industry far better than any wacky start-up, why didn't they do this before Tesla Motors came along?*

The answer? Inertia.

Why large manufacturing firms struggle to change what they do

Large companies that have been in business many years are a bit like the *Ever Given*, the ship that got wedged across the Suez

* To be fair, many of them had been making and selling hybrid vehicles for several years – but most were based around the premise that the internal combustion engine was the main power system, with a separate electric propulsion system for short urban journeys.

Canal. Big firms, like big ships, find it very hard to change direction quickly. They've invested vast amounts of time and money to fine-tune the operations of their giant factories, optimising their vast supply networks, gaining a sophisticated understanding of their customers and building up a workforce with very specific skills. Changing this intertwined system would be *very* expensive. And in the case of the automotive industry, we aren't just talking about all the firms involved in making and selling cars – there is the added weight of the fossil fuel industry, a significant proportion of which is focused on providing the fuel and lubricants for internal combustion engines.

This complexity can be viewed as a competitive advantage for big firms. If the basic requirement of success in the industry is running large factories requiring very specialised equipment and a large, highly skilled workforce, along with a dedicated network of specialist suppliers, not many new companies are going to want to try that. When the industry is about to change, though, cracks can appear in these barriers through which new competitors can sneak. For example, if the change is based on a new technology that the big firms don't yet have, then nimble new firms already in possession of that technology – and unencumbered by all those old manufacturing processes – may be able to jump in ahead of the older, slower firms. This is exactly what happened with Tesla.

But big firms do have some advantages over start-ups, one of which is their ability to do things on a grand scale. Large car firms have been deploying this advantage in recent years to try to regain ground lost to Tesla and their ilk. Realising that they needed to get products to the market as soon as possible, those companies couldn't risk not having anything available for their loyal customers while they spent the typical two to five years to get from concept model to showroom-ready.[10]

So, they started simple.

How do you design and make a new EV quickly?

To make this change as swift as possible, design engineers at the major car makers were given the fun task of converting some of their current ICE-powered vehicles into EVs. They hoicked out the parts that were required for internal combustion – the engine, gearbox, clutch, transmission, fuel system and so on – and shoehorned in an electric propulsion system. It wasn't elegant, but it resulted in EVs being quickly available from the major brands.

However, a car designed from scratch to be fully electric will be very different from an ICE vehicle. For one thing, they will be much simpler. A typical petrol or diesel car has around thirty thousand components; an EV has fewer than half that number.[11] An ICE car has some remarkably complex paraphernalia to get one to two tonnes of metal, plastic, rubber and glass moving along the road through co-ordinating thousands of little explosions.

You need systems to store and pump fuel, to convert the liquid fuel into vapour, to inject, compress and ignite the vapour at precisely the correct millisecond to fire pistons down cylinders. The straight-line motion of the pistons firing down the cylinder must be converted into rotational movement via a connecting rod and crankshaft. This rotation then needs to be channelled through a gearbox to ensure the appropriate level of torque and rotational speed is delivered to the car's wheels. Those thousands of explosions per minute require an intricate set of mechanical and electronic systems to ensure every precision-engineered valve, cam, rod, shaft and gear is in exactly the correct position at exactly the right time. Then there is a whole network of channels and pipes to ensure that all the moving parts are suitably lubricated, and another network to ensure that the substantial amount of heat generated by all these explosions is swiftly dissipated using radiators and fans.

The driver and passengers should be completely impervious to this mad symphony of explosive chemistry and dynamic mechanical engineering taking place a few inches from their seat – the driver just has to fill up the car with fuel from time to time, take it in for an annual service and get really cross when their phone seems to randomly refuse to connect to the stupidly over-complex in-car infotainment system.*

By comparison, the technology to propel an EV is almost laughably simple. All that mechanicalness is replaced by a battery pack (the biggest and heaviest component), at least one electric motor and some control devices to convert the battery's DC current to the AC required by the motors and to manage the use of the batteries. That's pretty much it. No complex transmission system, no clutch, no valves, no crankshafts, no fuel system, no lubrication system, no radiator, no . . . well, you get the picture.†

How EVs are changing how and where we make cars

As we know, the nature of a product dictates the nature of the factory that will make it, and the design of the supply chain and location of the factory will in turn be influenced by the technology at the heart of the product. A production line designed for assembling an ICE car is not the same as the one needed for an EV, and because EVs require fewer components, the supply network needed for these components is potentially going to be simpler. It might involve some of the same suppliers as before (you are still going to need wheels, brakes, seats), but also new ones, and as one of these components is the heavy battery pack,

* Or that might just be me.

† In the transition phase, it's amusing that some car companies still stick fake radiator grilles on the front of electric cars. The alternative is to have a smooth, radiator-less front end like the Tesla Model 3, which some find a bit 'wrong', likening it to Keanu Reeves in *The Matrix* when his mouth is removed.

you – or rather your battery suppliers – aren't going to want to have to shift these over a long distance.

The result of all this is that firms wanting to manufacture EVs have to either substantially reconfigure one of their existing ICE production factories or build a completely new factory. Both approaches have significant costs. Volkswagen, for example, spent well over one billion euros converting its factory in Zwickau from producing petrol-powered Golfs to fully electric ID.3s – this money was not just for all the physical conversions and new equipment needed in the factory, but also to retrain current workers or hire new workers able to deal with the specific processes required to safely manufacture EVs.*

But transition from ICE to EV is having an impact far wider than just the reconfiguring of car factories and their associated supply chains. The arrival of EVs is also changing the global structure of the car industry. China is now one of the biggest and fastest-growing centres of EV manufacturing – and EV purchasing. One Chinese company, BYD, produced just over three million of the fourteen million EVs sold worldwide in 2023;[12] and nearly 60 per cent of all EVs sold worldwide that year were bought in China.[13] Tesla's second-biggest single national market after the US? China.[14]

The current transformation of the industry, driven by the demand for cars that are less harmful to the planet, has been a long time coming. Many of the changes we are witnessing in the 2020s are not unique to this one corner of the manufacturing world. It is hard to find any part of the manufacturing world that is not having to change.

The truth is that manufacturing always has been, and always will be, in a state of transformation.

* At the time of writing, it is estimated that the global automotive industry employs fourteen million people.

How to be a successful manufacturer when 'everything changes and nothing stands still'*

Manufacturing systems work best when everything is stable, when everyone knows what they are doing and when interruptions and changes are kept to a minimum. Then, along comes a new idea, be it a new way of switching off kettles or propelling vehicles. The senior managers in manufacturing firms have to look at the technology and ask themselves: do we need to take this technology seriously? If so, what might it mean for what we make, how we make it and what we sell? And when there are multiple new technologies being developed with different advantages and disadvantages, how do we decide where to focus our efforts?

Coming up with useful answers to those questions rests, in part, on an understanding of how a *dominant design* emerges and is exploited.

How manufacturers build on a dominant design

Back in the early 2000s, in the far-gone days before video streaming services, consumer electronics manufacturers were developing different technologies to allow DVDs to store high-definition video.

But there was a problem.

These firms couldn't agree which was the best technology, and so products based on two rival – and incompatible – technologies were launched: HD-DVD and Blu-Ray. This duality made customers nervous. What would happen if they bought a fancy HD-DVD player and started building up their collection of HD-DVD movies, only to find that Blu-Ray goes on to become the most popular technology, with movies no longer being released on

* Plato, *Cratylus*, 402a.

HD-DVD? The same could also happen in reverse. Until this issue was resolved, sales of both technologies were very low. When the group of companies who had been pushing HD-DVD technology decided to call it a day, however, Blu-Ray became the dominant design, and the market exploded.[15]

Once there is a dominant design, firms can stop arguing about which technology is best and focus on developing, making and selling products based on that single design. But the emergence of the dominant design is often just the starting point for a cascade of activities that need to be completed before a product reaches the market.

The early days of the car industry illustrate this beautifully.

Back in the late nineteenth century, a whole load of things had to be in place for the market for this new-fangled smoke-belching, petrol-powered car to grow. Firstly, customers had to know that if they did buy one of these cars, suitable fuel would be available wherever and whenever needed. The whole infrastructure for finding, extracting, processing, standardising and selling petrol and lubricating oils suitable for cars had to be developed.[16] Then, a network of places where people could see, test-drive and actually buy cars had to be created. On top of this, as these early cars were quite unreliable, would-be customers needed to be reassured that, if things did go wrong, someone could fix them – customers needed confidence that there would be mechanics available with the right skills, equipment and supplies to maintain and repair cars.

And that's just the absolute basics. There was also the need for testing and approval systems to ensure all cars on the road were safe, smooth roads that had to be built and integrated with a network of fuel stations, and financial instruments to help people buy or lease these expensive assets, along with appropriately priced insurance; the emergency services also needed to know how to deal with crashes involving metal boxes containing highly volatile fuels moving at speed. We have had well over a century of refinement of this multi-layered internal-combustion-focused industry, one that has become deeply embedded throughout so many aspects of our lives and communities.

In industries as old, deeply rooted and complex as the automotive industry, sometimes someone else needs to step in to trigger any significant change. And given the major impacts that a major transformation of a whole manufacturing sector can have on local communities and national economies, that 'someone' is often the government.

'I'm from the government and I'm here to help'*

There are many things that governments can do to help manufacturing firms deal with big technological transitions,† and they broadly fall into three buckets. There are *supply side* issues – helping firms adopt and adapt to the new technology. There are *demand side* things – encouraging consumers to buy products based on the new technology. And then there is stuff we might call *dealing with unintended consequences* – like what to do for local communities when the arrival of a new technology leads to job losses.

Let's look at some of the contents of each of those buckets in a bit more detail.

On the supply side, one of the simplest tools, if a bit blunt, is to just mandate that a thing must stop, forcing manufacturing firms to find a way to do things differently. For example, in late 2020, the UK government announced, 'Sales of new petrol and diesel

* 'The nine most terrifying words in the English language,' according to US President Ronald Reagan.

† I use the word 'can' deliberately as I don't want to get us dragged into a discussion about political ideology and role of the government versus the role of markets.

cars to end in the UK by 2030' – no ifs, no buts.[17] While such policies do provide an unambiguous target for automotive firms to work to,* it would be a bit rich to just leave them to get on with sorting that out on their own. We know that the car industry is incredibly complex, employing millions of people, using techniques developed and refined for over a century. Without a bit of practical help, those firms are really going to struggle to change.

But what does that practical help look like?

How governments can help de-risk new technologies

In the disappointingly cool summer of 2023, I drove due west into the UK's industrial heartland to see what government support for changing the UK automotive industry looks like in action.

Exiting the motorway, I wiggled through leafy West Midlands suburbia that gradually blended into the Warwick University campus. Despite being told in advance that it was OK to do so, I still felt guilty as I turned the opposite way to the 'All Traffic This Way' sign – I mimed an apology to cyclists and pedestrians – to drive along a pedestrianised street towards the Warwick Manufacturing Group's (WMG) International Manufacturing Centre.

'It's probably easier if we don't watch you,' laughed the helpful receptionist as I demonstrated my incompetence at trying to log in via a touchscreen. Having scraped through that test, I met my host Robin, who took me to lunch before talking me through what happens in these buildings.

WMG forms one part of a multi-hundred-million-pound UK government initiative designed to support industrial transformation, to provide that 'extra oomph'. Called the High Value Manufacturing Catapult, it is a network of seven centres spread across the UK, each crammed with highly skilled engineers and the latest manufacturing

* Unless there's an election looming and you get a bit nervous: in September 2023, UK PM Rishi Sunak delayed the ban to 2035.

technologies. Each centre has one or more specific areas of expertise, be it a particular production process (such as making things using composite materials) or a sector (such as renewable energy generation). WMG has particular expertise in helping firms speed up their transition to electrical power systems.

Centres like these have a common purpose. They try to address two specific problems that hinder the use of new technologies by manufacturing firms. The first relates to the process of *industrial scale-up* – working out how something that works on a lab bench can be reliably, consistently and cost-effectively made in high numbers in a factory.

The building just across the road, the Energy Innovation Centre, along with the nearby Battery Industrialisation Centre, are examples of government-supported activities to help battery and EV manufacturing firms iron out any problems with production of batteries at gigafactory scale.

Battery and EV manufacturers need to be supremely confident that every cell of every module of every battery pack in every vehicle is going to work as intended (and not cause the vehicle and its occupants to be consumed in a fireball arising from a tiny defect that occurred during production). To do that, every aspect of the process of making a battery must be characterised and tested in exhausting detail; repeatability of process and consistency of product are paramount. Manufacturers know that if one defective battery were to leave the factory and end up in someone's EV, the results could be terminal not only for the vehicle's occupants but for the manufacturing business as well.

You might be asking why manufacturers can't just do such development and testing themselves. Why should the government spend taxes paid by us hard-working citizens to help private companies sort out their problems? The logic is that by making such facilities available to all firms, more firms are likely to adopt the new technologies, and we are more likely to have a rapid adoption of new battery technologies to accelerate the shift to EVs, which benefits the whole country . . . and the planet.

And if there are to be any 'thermal events' with EV batteries, I'd rather they happened in the controlled environment of a lab in Warwick than when I am whizzing back home along the motorway wondering whether it is my imagination or I really can smell burning.

How governments can help manufacturing firms start using new technologies

The second problem WMG-like centres are seeking to address relates to what industrial policy geeks refer to as *technology diffusion*. Even if a technology has been around for a while, some manufacturing firms may be reluctant to use it in their factories. This can be clearly seen in the slow adoption of computerised automation for various manufacturing activities, especially those within smaller manufacturing firms. This reluctance to adopt new technologies can lead to some big problems.

Over recent years, economists have been scratching their heads about why smaller UK manufacturing firms aren't as productive as their international peers.[18] Small manufacturing firms in other countries seem to be becoming more productive – they make more things per hour – while UK firms' performance remains obstinately lower. After much debate, one conclusion has been reached: UK manufacturing firms would be more productive if they adopted more digital technologies to improve the efficiency of their factories.

So why aren't they doing this?

Surveys have revealed many reasons.[19] One is that managers seem unsure whether their workers have the skills needed to use these new technologies; another is that managers are often unsure how difficult (read: expensive) it might be for these new technologies to be adapted for the specific needs of their factories. Those are entirely sensible concerns. Why would you risk buying an expensive new bit of kit unless you were absolutely sure you knew how to use it, and you knew it could definitely improve the performance of your business?

Governments can help small factory owners with both of these problems. The government can fund training courses to upskill people working in manufacturing firms so that they can adopt new manufacturing technologies, and demonstration facilities can be built to allow manufacturing managers to come and have a play with the new technologies, to get a practical sense of what such technologies might be able to do for their firms.

A great example of such a demonstrator facility can be seen in a somewhat sinisterly named facility located sixty or so miles north of Warwick: the Omnifactory.*

It is a great piece of industrial theatre. I'm standing in the control room of the Omnifactory, located in a shiny gold building on the edge of the University of Nottingham's Jubilee Campus. Spread around the walls are large computer screens showing a simulation of different industrial robots and mobile platforms configured into different layouts to assemble various types of large aircraft components. With a click of the mouse, the movement of each robot and platform – and the whole production system – can be digitally simulated to show how things can be swiftly and automatically rearranged to make different products. Along with the other visitors, I nod politely as this plays out before us on the screens.

* Every time I read, write or say 'Omnifactory', I think of the terrifying self-aware, self-learning and highly destructive 'Omnidroid' from the Pixar movie *The Incredibles*. I keep meaning to ask colleagues at Nottingham University whether their marketing team have seen this movie.

Then Svetan, the head of the lab, smiles as a button is pressed and the giant opaque sheet of glass at one end of the room turns transparent and . . . *voilà*. There is the actual factory, massive robots and all, its layout being automatically reconfigured to replicate exactly what had just been shown on the screen. This is a prototype of one model of the factory of the future: a place that can make any product. The role of the Omnifactory is to show manufacturing firms the possibilities and effectiveness of the latest automation and digital technologies.

I'm impressed, though I can't help but feel a slight sense of unease. There is something about the scale and movement of the robots that reminds me a little too much of one of the early scenes from the movie *Avatar*.

What confronts me upstairs, though, triggers quite a different reaction. In a long cabinet with transparent sides, little white Kuka-branded robots on movable platforms are waving their short arms around, pretending to perform some sort of assembly task. It's almost as though they are the tiny offspring of the monster-sized robots in the lab below, still learning what to do. This is a prototype 'mini-Omnifactory', designed to show smaller manufacturing firms with more modest budgets how they too could benefit.

WMG and the Omnifactory are examples of ways in which taxpayers' money can be used to help manufacturing firms deal with the arrival of new technologies. Both aim to help reduce the risk and uncertainty for the manufacturing firms pondering the adoption of a new technology. It is about the government helping the supply side – the manufacturing firms – by helping them gain confidence and reduce risk.

There are also, however, things that governments can do on the other side of the equation, the demand side, to kick-start markets for the new technologies.

How do governments help stimulate demand for new products?

Sometimes, no one wants to go first, to be the first customer to take the plunge on new technology. Into the breach can step the government. By acting as an early customer, they can show others how a technology can be used, and also provide those early innovators with much-needed income from initial sales.

As we shall see in the next chapter, governments have played a particularly pivotal role as a customer in stimulating the early market for computers. The example of the dawn of the electronic computing industry in the US and UK shows this clearly: IBM's first forays into electronic computing were de-risked by the US government spending hundreds of millions of dollars on their products. As we shall also see, the UK government's decisions *not* to do so stymied the potential of one computer company that some believed could have rivalled IBM.[20]

Governments can also do things that encourage you and me to buy products based on new technologies. This has been one of the most visible forms of government support for the transition from ICE cars to EVs.[21] Subsidies can be provided to reduce the price paid by the customer for a new EV, which is one of the key reasons sales of EVs grew so vertiginously in China in the early 2020s. To address the issue of 'range anxiety' that stymied sales of the earlier generation of EVs, governments can also invest directly or indirectly in the network of charging stations.

On the other hand, governments can also do stick as well as carrot, disincentivising the purchase and usage of certain products. To stimulate EV sales, they can make it difficult for people to buy and run fossil-fuelled vehicles. From the 'Ultra Low Emission Zones' being rolled out in cities around the world and increasingly stringent regulations on ICE vehicle emissions, to just cranking up taxes to make petrol and diesel really expensive: all can play a part in nudging consumers towards low- or zero-emission vehicles.

What about dealing with those 'unintended consequences' of a manufacturing transformation?

With any major manufacturing transition, there will be winners and losers. The adoption of technologies such as Honoré Blanc's interchangeable parts led to improvements in productivity, but also cost many highly skilled gunsmiths their jobs. It's the same today. The ongoing adoption of automation in manufacturing almost always leads to job losses, even if some workers can be retrained and new jobs created. This is especially the case as the industry moves from the relative complexity of ICE vehicles to the simplicity of EVs, leading to changes across the whole automotive supply chain. Compounding this, like the rest of the economy, the manufacturing world is seeing the impact of AI on its workforce, with some processes requiring less and less human input.

While companies can do many things to help support workers dealing with these transitions, some things will be beyond their control – or at least beyond their budgets. Again, this is where governments can step in by providing funds for retraining, helping develop new employment opportunities and providing welfare services to local communities disrupted by unstoppable waves of technological or economic change.

The mining town of Orgreave in South Yorkshire is symbolic of one of the most painful periods in recent UK industrial history, the backdrop of which was shaped by one such unstoppable wave. In the 1980s, this was the scene of violent clashes between striking coalminers and police arising from a bitter set of disputes facing the oft-troubled, slowly transforming UK steel and coal industries. In brief, the Orgreave mine was eventually shut down by the government, resulting in hundreds of job losses and the local community being decimated.

Fast forward to the 2000s, however, and thanks to direct local and national government help, Orgreave has a new lease of life. It is now home to the Advanced Manufacturing Research Centre (AMRC), one of those facilities like WMG and the Omnifactory for de-risking and demonstrating new production technologies. As an endorsement of its success, clustered around this centre are manufacturing facilities owned by firms including Rolls-Royce, Boeing and McLaren. The AMRC also runs a facility for training apprentices who can move into high-skill, well-paid manufacturing jobs.

Despite Ronald Reagan's comment at the start of this section, it is pretty clear that governments can and do play a critical role in supporting manufacturing firms coping with a transformative industrial change. They can also play a key role in dealing with other unintended consequences of such transitions.

Not every innovation is radical or disruptive or transformative

Much of what I've covered so far in this chapter has focused on examples of substantial change to a product or a process, change that is going to be highly disruptive to current manufacturing activities. But there is another type of change we need to cover in this chapter. It affects all manufacturing firms, is hugely important, yet is regarded by some as a bit Cinderella-ish.

To explain, let me take you to a room filled with three hundred first-year engineering undergraduates in front of whom I have just plonked a car-battery-sized 1980s mobile phone.

'Why'd they make it so big?' asks one student.

I work with some of the smartest people in the world, but sometimes they don't seem quite as on-the-ball as I anticipate. To be fair, no one is really at their best at 9 a.m. in a stuffy lecture theatre on a leaden-skyed Tuesday in November. And maybe that student's question is not an unreasonable one. The first mobile phones were indeed 'so big', but they were a remarkable feat of engineering, drawing upon the latest technologies available at the time.

The early phones were also ludicrously expensive, had very short battery lives and had limited places from where you could actually pick up a signal and make a call – unsurprisingly, the number of customers was tiny. But then, slowly, a cycle of events began that grew the market from a handful of rich, tech-loving nerds and show-offs to there being over twice as many mobile phones in the world as there are people.[22]

How did that happen?

Why people shouting 'Hello? HELLO? . . . WHAT? . . . Yeah, I'm . . . I'M ON A TRAIN . . .' is important for innovation

As the benefits of mobile phones became apparent, more people started to buy them, increasing their visibility to potential customers – even if this did lead us to having to listen to annoying shouted conversations on trains. This in turn meant that more investment went into improving functionality and reducing cost to make them available to an even bigger pool of customers, and these steps became a cycle. As performance continued to rocket upwards – as seen in the evolution from 1G to 5G, for example – costs continued to come down,* and so, in a relatively short period, we ended up in a world where mobile phones outnumber people.

* Yes, some smartphones are ludicrously expensive but that's not directly linked only to their performance. There are a whole load of branding and marketing issues at work that result in some firms being able to find customers who willingly pay a substantial premium.

This is a great illustration of this other type of change that happens in manufacturing: *incremental innovation*. There's that first phase when a new thing is discovered or invented, which is then followed by step-by-step improvements being made to 'scale up' the making of products based on these ideas. These iterative steps could relate to how the thing is made (Have we designed the process to be as efficient as possible to make consistently reliable products?), or they could focus on ensuring the product can be used by customers (Have we provided a cell phone network that allows customers to use their phone in enough locations?). The improvements could also relate to the technology itself (and yes, making phones not 'so big' was part of that).

This process of making incremental, continuous improvements is something that manufacturing firms will do throughout the whole period during which a product is made. Everything we looked at in the first half of this book – the making, the moving, the responding to customer demand – is constantly being tweaked and refined by manufacturers.

Managing continuous, relentless change

As products are continually refined, we often forget (or are blissfully unaware of) what manufacturers have had to do to develop them to their current levels of reliability, cost and quality.

In a corner of my office sits a big plastic box, roughly the size of family holiday suitcase, containing a slice of telephonic history. Specifically, it holds examples of some of the most popular (and weirdest) mobile phones sold between the mid-1980s and 2010s.*

* The car-battery-sized one I mentioned earlier is not actually in the box – that one has been repurposed as a very effective office doorstop for use on windy summer days.

Taking each one out, I lay them in a line in chronological order to see their size shrink from shoebox- to matchbox-size around the mid-2000s, and then start to grow again as bigger screens were needed for accessing online content rather than just talking and texting. Roughly in the middle of the line is a Nokia 3310 – this is the phone that I can hold up at any talk or event, ask the audience if they ever owned a phone like this and marvel at how the majority of hands will nearly always be raised. They'll then remember the wonders of a phone that had a battery life that could be measured in weeks, and a concrete-like robustness far removed from the fragility of today's smartphones.* And, of course, they'll revel in the nostalgia of that wonderful Nokia phone-based game, *Snake*.†

This line-up by size gives a very superficial version of the story. Yes, phones got smaller in response to customer preferences, and engineers worked their magic to shoehorn components into ever tighter spaces, but the technologies that underpin mobile communications were also developing apace and increasing in complexity, meaning engineers had to continue streamlining their work. Then, of course, phones ballooned in functionality with the development of the smartphone, and that took us from talk and text to full mobile internet access almost everywhere on earth.

Every one of the hundreds of components and subsystems that are brought together to make a smartphone had not only to be invented, but to go through a relentless programme of development – as with the EV batteries – to ensure that they could be manufactured reliably and cost-effectively in batches of millions. Billions of dollars of investment were made in factories around the

* This was an era when products like shock-proof cases and screen protectors were only required for military applications, when your phone needed protection from explosions, rather than protection from daily life.

† You can, of course, bathe in waves of nostalgia via emulators – just type 'Nokia Snake' into Google Play or the App Store.

world to manufacture and relentlessly, continuously improve some of the most complex, miniaturised electronic systems ever made.

As customers, we don't see any of that. We just look at a beautiful, shiny device that has become an indispensable tool of modern life, and grumble when it can't stream high-definition video as our train enters a tunnel. I keep my collection of old phones to remind myself of this largely invisible and often forgotten aspect of manufacturing change.

We can now see that there is a cycle common to almost all manufacturing transformations. It starts with the development of a new technology that best fits a market need: the dominant design. Once we have that, manufacturers target their efforts on making myriad small improvements to the product and the way it is made, so that it can be produced at scale, defect-free, at an appropriate cost. And, in the case of products that require some kind of 'consumable' to make them do something useful for the customer – be it petrol, electricity, ink, razor blades, bullets – they also help create the necessary infrastructure to provide those consumables at the right quality, cost and location.

But before we draw this chapter to a close, I want to answer a few questions that might have been bugging you for a while.

Who is pushing for all this relentless, exhausting, endless change? Why the need for constant improvement? Why not be satisfied with how things are now?

What's wrong with *That'll do*?

Why do we keep innovating?

Japanese professor Noriaki Kano came up with an elegantly simple insight that answers these questions.

Imagine you are back in the early days of the car industry. If you achieved the remarkable engineering feat of making a vehicle that could consistently travel more than about twenty miles, you were pretty sure of attracting some loyal customers. Through the processes described earlier in this chapter, the reliability and range of all cars improved over time, and boasting that your car could travel over twenty miles soon no longer differentiated you. You needed to either improve your performance against that measure or find some other measure of relative performance – you could, say, manufacture a car that was able to start more easily on a cold morning, or one that could pass more smoothly over bumps. There were also other things that you, as a manufacturer, could highlight, things that might delight the customer – for example, you could talk of the comfort of the seats, or the existence of a roof over the driver and passengers. However, people wouldn't buy a car if it had a roof but wouldn't start in the morning or broke down every few miles.

Customers have specific requirements which Kano categorised as *basic*, *performance* and *delighter* features. And here's the really important bit that underpins all that faffing around with those continuous, incremental improvements: those customer requirements are not static.

Just think about it for a moment. If you were considering buying a car today, the relative reliability of different brands of cars starting on cold mornings would probably not be a major factor; I also suspect that the existence of a roof would not have a big influence on your purchasing decision. What happens over time is that customers expect more and more as standard. *Delighter* becomes *performance* (*Wow, that car has doors!* becomes *Hmm, this car has doors that can be locked whereas that one doesn't*). Eventually, it reaches a point where no one talks about doors at all – it is a *basic* feature of all cars, and any salesperson who wants to draw your attention to quality of the doors might be trying to divert your attention away from some less desirable feature of the car.

Kano neatly encapsulated why manufacturing firms are – and in fact have to be – in a constant state of flux. They must respond to the relentless need to differentiate their product from others. Each time one firm pushes ahead with some increase in performance or new feature, all the other firms are forced to do the same to remain competitive. This levels out the competition, so firms then have to find some other way of improving, of making their products 'better' than their competitors'.

And that is why manufacturers have to keep innovating.

Linking together all the stories we've seen in this and earlier chapters, you can now see why manufacturing managers are often so frazzled. As I hope is now super clear to you, manufacturing works best when everything is stable: predictable demand, reliable supplies, smoothly running machines, happy workers. But nothing stays stable – customers are finicky, suppliers can't always supply, machines break down and workers . . . well, are people. Then, on top of that, you have all those disruptions that span the immediate crises (pandemic, wars, etc.) to the slower-burn changes (the arrival of a new technology or a change in social attitudes). And sometimes, you get the hideous combination of all those things at once.

This chapter has shown that big changes can be made within the manufacturing world, and that these changes can impact almost everyone. And we've also seen that managing this is not easy, but it can be done.

I'm back standing on the shop floor of the Strix factory on the Isle of Man, mesmerised by the shiny chains of part-finished blades ker-chunking their way out of steampunk-esque machines. This feels like 'proper' manufacturing. Oily, noisy, visceral, precise, profitable, sophisticated, metal-bashing manufacturing. This is analogue manufacturing at its best.

However, as everything else in our lives goes ever more digital, I can't help but think: What will happen to factories like this?

6: Connect

How digitalisation transformed – and continues to transform – the manufacturing world

As I slid the key into my front door lock, my peripheral vision caught something that felt wrong. Down the road through the late winter morning gloom, I could see a bicycle lying on the ground, a hoodie-clad figure slightly hunched before a neighbour's door. I turned and walked towards him, pulse slightly elevated, breath condensing. Getting closer, I was shocked when I saw what he was doing. He was delivering someone their morning newspaper.

I couldn't remember the last time I had seen a newspaper being delivered on our street. In fact, I struggled to remember the last time I had bought a physical newspaper. While I know I'm a sample of one, and a huge number of people *do* still avidly consume their news in paper form, the decline of the newsprint industry over recent years has been dramatic. Just look at these numbers for the previous couple of decades:

> Annual spend of UK households on newspapers has dropped from over £9 billion in 2005 to £3 billion in 2021;
>
> In 2000, sixteen paid-for newspapers in the UK had a combined circulation of over twenty million. In 2020, the number had dropped by two-thirds;
>
> Demand for paper for newsprint fell from two and a half million tonnes to just over one million between 2000 and 2017.[1]

What happened? The shortest summary is that we shifted from moving atoms to moving electrons to deliver our news. The development of the internet – especially when accessible via our phones – changed the whole way we consume content. Why go through the faff of arranging for folded sheets of printed-upon dried wood pulp to be delivered to your home or bought at the newsagent when you can have the same content delivered instantly to the phone in your hand or the computer on your desk?

At the root of the transformation of the newsprint industry is digitalisation, a set of processes for capturing, analysing and communicating data in digital form. These mundane-sounding processes have transformed almost every corner of the manufacturing world and, in turn, almost every aspect of our lives.

How this happened is a remarkable tale of revolutions, mad inventors, cake and the building of two giant machines. I want to share this story with you to show how digitalisation has had such an impact on how things are produced, moved and consumed by us today. I also want to share with you some of the amazing yet quite scary things that could happen to us if manufacturing continues to accelerate along this digital path. To begin this tale, I'd like to set three scenes before you:

Scene 1

It's late Sunday evening. I've just finished packing for a trip to China to attend a conference and I now can't be bothered to cook. I reach for my phone and swipe to a collection of apps grouped into a folder I've labelled 'Lazy'. Tapping one that recently emailed me a discount code, I assemble and order a pizza from a popular local store. In less than half an hour, Mo arrives at my door via electric scooter and hands me a warm cardboard box containing exactly what I ordered. Every step of my pizza's production and Mo's journey was trackable via my phone.

Scene 2

The following evening, I am in seat 43A on an Airbus A350-900 drinking coffee from a paper cup, peering out over the North Sea at the start of the long arc over Russia to China. Out the window to my left side roars one of two Rolls-Royce Trent engines. These engines provide the thrust needed to carry the hundreds of tonnes of aircraft, fuel, me, the other passengers and all our luggage smoothly, high and fast above the frozen Siberian tundra. For the ten hours of the flight, the high-pressure core of each engine spins without a glitch at around 12,500 rpm at a temperature of 1,700°C. The uneventful journey concludes as we land in Beijing in time for lunch.

Scene 3

A few days later, I'm sitting in a minibus passing through the gates of a sparkling-white new factory on the outskirts of the eastern Chinese city of Ningbo. Aside from its whiteness, its other obvious feature is size: the whole factory site covers eight square kilometres. We draw up to the main entrance, where a giant Chinese national flag flutters on a pole just slightly higher than the neighbouring corporate ones. Standing outside the imposing glass doors, clad in black T-shirts, our hosts are waiting to give us a tour of one of the newest and most advanced EV factories in the world. This is not just a factory: this is a smart factory.[2]

What connects these scenes? They all illustrate the commercially successful use of the most valuable commodity for any manufacturing business today: digitalised data.

Thanks to technological developments with roots going back several centuries, manufacturing firms today have access to eye-poppingly large and ever increasing amounts of this commodity. As we know, the problem many firms face is knowing what to do with this constantly growing mountain of data – some, however, know exactly what can be done, and looking at pizza delivery, jet-engine maintenance and car assembly is revealing.

First, pizza

Why pizza? Because it was one of the first food products to be 'internet-enabled'.

Pizza-like food has a very long history,[3] but the version you and I are most familiar with emerged in Naples – rather appropriately for this book – at around the same time as the start of the First Industrial Revolution. Waves of emigration from Italy slowly spread pizza across Europe and the Atlantic, and as the US economy boomed through the 1950s, two innovations expanded pizza consumption from restaurant to home: freezing (first marketed in 1957)[4] and delivering (first arriving in 1960).[5]

Fast forward to 1994, and the world changed once more: 'PizzaNet' arrived.* This allowed anyone with a PC and a modem, 'for the first time, to electronically order pizza delivery from their local Pizza Hut restaurant via the worldwide Internet'.[6] Initially only available to people in Santa Cruz, California, it is reckoned to be the first online food delivery service.†

Through the 1990s into the early 2000s, online food ordering grew in popularity, but we had to endure the glacial reassembling of grainy images of limited pizza options as the data trickled into our PCs. After clicking 'Order this pizza', there was a nervous wait as we wondered whether our order had actually gone through, payment had been made (or had we just passed our credit card details to some online scammer) and anything was really going to arrive at our front door.

* An exciting start-up called Amazon also launched that year, and could be argued to have had a slightly wider impact on the world.

† And for nostalgia's sake I've put an early version of Pizza Hut's online ordering website on this book's website, www.timminshall.com.

Today, we have near instantaneous access to food delivery via our phones, with ludicrous levels of choice being available via a system that ensures financial security as well as near real-time traceability of the whole production and delivery process.

Getting from PizzaNet to the likes of Deliveroo and Uber Eats has required vast amounts of investment and entrepreneurship. Someone had to develop a way of capturing, analysing and co-ordinating the flow of information to and from suppliers (the restaurants), deliverers (the drivers and riders) and the customers. Someone had to build and maintain a trusted, seamless and easily usable payment system that worked for all three parties involved. Then, there needed to be a way to curate the vast range of options available, so that we were swiftly guided to what we want but ensure these were synchronised with what the suppliers could (or wanted to) provide at a given moment. Finally, each stage of this process had to be visible to both producer and consumer.

All this had to be accessible – almost instantly – to customers, most of whom want to search and select via a few taps on a small touchscreen.[7]

Managing flying data capture systems

Returning to seat 43A on that flight to China, there are several things in common between what we have learned about pizza delivery and what we are about to explore in regard to the operation of the engines that power the majority of the world's commercial aircraft.

Jet engines have now become multimillion-dollar flying data-capture systems. Sensors embedded in the two Rolls-Royce engines hanging on the wings of the Airbus A350 were, throughout the flight, generating, storing and transmitting data back to engineers in operations rooms back in Europe.[8] Those engineers were staring at screens displaying data coming in from every connected engine in use around the world. With this data, they can spot tiny signals that give early warning of any required maintenance a little earlier than planned, or get an early hint of something that might need tweaking or repairing. With this information, they can make sure that an appropriately skilled engineer with the right part or tool is waiting when the plane lands to fix the specific problem. In the latest versions of this system, those same engineers can also send tailored requests to the engines mid-flight, using those sensors to take a closer look at some specific issue they have spotted.[9]

All that was going on as I stared vacantly out of the window and wondered which way to go with the impending chicken-or-pasta question as my nose sensed the start of the food service.

Digitalisation makes factories smart

A few days later, I walked into the cool, sleek atrium at the centre of Zeekr's massive new electric vehicle production site in Ningbo, China.* Dead ahead on a raised platform was just one car, a bright orange Model 001. We waited politely as giant screens behind the car played a promotional video to a thumping techno soundtrack while slowly sliding to reveal another space. There, samples of components, materials and paintwork were on display. This was all beautifully presented and all very shiny, just as you might see in any modern car showroom.

But over to the left was something I found much more interesting, and which explained the reason for our visit.

It was another giant screen but this one was displaying images and live data from activities across the massive production site. This was possible as the factory has its own dedicated, private 5G network. With this, the Zeekr managers have a continuous stream

* At the time of writing, Zeekr was part of the Geely group of companies that also own the Volvo, Proton, Lotus and Polestar brands.

of data from thousands of sensors on the hundreds of machine tools, robots, automated guided vehicles (AGVs),* conveyors, storage areas and energy systems across the vast site. These sensors cover the entire manufacturing process, from the casting and machining of major structural components, through the assembly of the fifteen thousand or so individual mechanical and electronic components, and finally to the embedding of the millions of lines of software code needed to bring the car to life.[10]

With this data and high-speed analysis, managers can almost instantly spot glitches in the smooth operation – like a malfunctioning robot – and quickly reroute activities while the glitch is dealt with. Even better, as with those jet engines, they can spot problems before they get too serious and do pre-emptive maintenance on the problematic machines.

And the managers also have a stream of data coming in from the customer side. As with many new car companies, Zeekr is very much an online business. They've made substantial investments to build an internet-based sales channel and put particular emphasis on customers connecting with Zeekr via their mobile phones. With this data, Zeekr – and their hundreds of component suppliers – are better prepared to respond to ever changing customer demand. This allows them to use that tricky strategy described earlier of *demand chasing*: the tight coupling of predicted demand and making of exactly the right products.

Zeekr managers now have a digital thread of data that connects each customer and their unique preferences to the intricate manufacturing system that builds each car. That connection will remain throughout the car's life, allowing software upgrades to be delivered and operational data to be captured to ensure the car is serviced when needed. And that data also provides the manufacturer with

* With this network, the AGVs can use SLAM (Simultaneous Location and Mapping) to work out where they are, where they need to go and what is going on in the immediate vicinity without any human input.

valuable insights on how to improve the design of the next generation of cars.

These examples illustrate some of the ways digitalisation has impacted manufacturing. The Zeekr example in particular shows how digitalisation has transformed the *making* of the product, the pizza story shows how it has changed the *moving* of the product, and the jet engine the *use* of the product. And all three illustrate how digitalisation is allowing ever more seamless integration.

Digital technologies are doing the same for almost all manufacturing sectors, but as we will shortly see, impressive though the speed and scale of these changes are, what we are witnessing now is just the latest stage of a manufacturing metamorphosis that started two and a half centuries ago.

And it is far from over.

In late summer 2023, aided by a jug of extra-strong coffee, I sat down to research and write a simple summary of the current and emerging trends in digitalised manufacturing for this chapter.

Just minutes later I was in that state described by internet scholar Manuel Castells as 'informed bewilderment'.[11] There was just too much hand-wavy, waffly vagueness; so many new terms, acronyms and gushing hyperbolic visions of the future. Sighing, I realised that I first had to go back to the basics. I needed to understand how 'digital-ness' had arisen and merged with the processes of making, moving and consuming over the three previous industrial revolutions. I closed my laptop and went for a walk to clear my head.

When I returned, my starting point was a faint memory from my school days about the use – and resistance to the use – of some kind of punched cards to control looms a couple of centuries ago. However, I have to admit I was a little vague on the details of how we moved from automating and smashing up looms in revolution-era France to 5G-enabled smart factories and Snoop Dogg promoting Just Eat.

To fill in some of the blanks, we need to return to the first industrial revolution.

Digitalisation and the First Industrial Revolution

In the early 1800s, French weaver Joseph Jacquard was facing a problem. The growing market for patterned textiles and the swelling number of middle-class consumers was putting pressure on the availability of highly skilled workers to set up and run the looms. To address this problem, Jacquard developed a way of storing information about textile patterns on sets of punched cards. The presence or absence of holes in those cards directly controlled the raising or lowering of threads in a loom. This allowed any desired pattern to be obtained with minimal human intervention from any loom equipped with Jacquard's technology. In common with the arrival of so many new manufacturing techniques – like Blanc's interchangeable parts for weapons – this made many skilled workers extremely angry as it reduced their opportunities for well-paid work.[12]

Jacquard's use of punch card technology, however, had demonstrated the importance of something for manufacturing that went way beyond the control of looms.

Charles Babbage's 'analytical engine' of the 1830s is credited as the first general purpose computing machine – a device capable of performing a range of arithmetic and logic operations without human intervention.* It drew upon Babbage's earlier work developing a

* Babbage was also a great enthusiast for manufacturing and produced a classic book on the subject: Babbage, C. (1832). *On the Economy of Machinery and Manufactures*. His name is given to the 'Babbage principle' that describes the

'difference engine' (very impressive, but more calculator than computer) and the work of others, including Joseph Jacquard. This amazing device was fed instructions encoded in the presence or absence of holes in punched cards, but Babbage needed a way of giving the machine the 'recipe' – or programme – to follow. He required someone to write the software to be encoded on those punched cards. English aristocrat Ada Lovelace collaborated with Babbage and wrote the world's first computer program.[13]

This period also saw several advances towards transmitting data over large distances, an essential if rather small early step towards today's internet.

One of the initially most successful technologies for doing so started in a kitchen.

With his intended career as a clergyman derailed by the French revolution, Claude Chappe made a rather abrupt career change and started working on the development of long-distance communication technologies. His early attempts at transmitting simple messages were based on giving items of kitchenware a good whack using modified clocks (yes, really).[14] He had more success when he switched from a sound- to a sight-based approach. His optical system was based on synchronising pivoting black and white wooden panels with clocks, telescopes and a code book,* and it worked surprisingly well.[15]

Except when it was dark. Or foggy.

To provide a less daylight- and weather-dependent system, many inventors had been working on various electricity-based ideas.[16] One of the most successful was developed by artist turned inventor

---

advantages of matching different tasks to workers with specific skills. And so it is wonderfully fitting that the Institute for Manufacturing in Cambridge is located on Charles Babbage Road.

* Initially called the *tachygraphe* (Greek for 'fast writer') before switching to the now more familar *télégraphe* – '*far writer*'.

Samuel Morse. Morse developed not only the eponymous dot-dash coding system but also a way of transmitting messages over long distances along copper cables.[17]

Digitalisation and the Second Industrial Revolution

The second revolution wasn't just about increased scale in terms of the quantity of products; it was also about geographic scale. The availability of low-cost steel enabled the growth of the railways, allowing raw materials and finished goods to be transported over longer distances; the roll-out of electrification meant that energy for powering machines could be accessed remotely from the source of its generation. Factories could be established far from sources of raw materials or power and far from the markets for their products, and communication was enabled by the widespread use of the telegraph (once Morse's technology became the dominant design).

So, part of this second revolution was about technologies that had been around for a while being deployed at much increased scale.

But this period also earned its revolutionary badging from the scale and speed of the adoption of a whole raft of new ideas that we now call 'information and communication technologies'. Just to give a flavour of what was going on, here are a few.

In communication, we moved from dots and dashes to live, synchronous voice transmission over those same copper cables thanks to, amongst others, Alexander Graham Bell.[18] We then broke free of the wires thanks to the work of Italian aristocrat Guglielmo Giovanni Maria Marconi.* Marconi's technology

* Some smart alec provided this explanation for those seeking to understand this new-fangled wireless compared to wired telegraph: 'Well, if you had a very long cat, reaching from New York to Albany, and you trod on its tail in New York, it would throw out a wail in Albany. That's telegraphy; and wireless is precisely the same thing without the cat.' This quote has been attributed to Albert Einstein. Sadly I couldn't find any evidence that he actually said it.

allowed lower-cost transmission of messages between two points (no need for all those copper cables slung between telegraph poles or laid underground or undersea) – it also enabled mobile transmission, which proved to be especially useful for shipping.[19]

As manufacturing organisations grew, new approaches to capturing, processing and analysing data were needed. One key development to address this came from Herman Hollerith, a German-American statistician and inventor, who was drafted in by the US government to figure out how to capture information about their rapidly growing population. Drawing upon ideas sparked during the First Industrial Revolution, he developed a mechanical approach for automating the census-related data processing. His device – yet another one with links back to Jacquard's punched-card-controlled loom – was hugely successful. In fact, his system was *so* successful that it served as a core technology for what would eventually become one of the world's biggest computer manufacturing companies, IBM.

Digitalisation and the Third Industrial Revolution

By the end of the Second World War, manufacturing output had been cranked up to staggering levels,[20] and, as often happens in periods of crisis, a 'needs must' mindset allowed engineers to crash through many of the usual roadblocks to innovation. In the face of existential crises, those barriers to innovation discussed in Chapter 5 tend to dissolve away.* A key discovery of this period was the power of computers (both analogue and electro-mechanical) to solve head-achingly complex problems – like cracking the German military's Enigma code or making an atomic bomb.

In response to the economic slump that often follows a major conflict, manufacturing firms in the late 1940s sought new

* For example, penicillin, radar, synthetic rubber and the jet engine were all catapulted into widespread use during that conflict.

opportunities. Some saw the potential of those new-fangled computing machines to help make their businesses more efficient, and I was delighted to learn that one of the first firms to do this successfully wasn't in the field of rocketry, high finance or drug development. It was in the business of making and selling cakes. I learned about this thanks to a random seating allocation at a Cambridge college dinner back in the early 2000s.

As I drew my chair closer to the long table, the person on my right turned to introduce himself.

'I'm Maurice Wilkes.'

My brain struggled to process what had just been heard. Maurice Wilkes was one of the most influential figures in the early development of electronic computers. I had just sat down next to the computing equivalent of Leonardo da Vinci or Albert Einstein. As my composure recovered and the dinner progressed, I learned that the connection between my dinner companion, electronic computers and commercial baking was quite simple.

In the late 1940s, Maurice led the team that designed and built EDSAC,* one of the world's first stored programme computers – a fully functional electronic descendant of Babbage's mechanical analytical engine. EDSAC was initially used as a research tool by Maurice and his colleagues at Cambridge University.

Around the same time, two senior managers at one of the UK's most successful catering firms, J. Lyons and Co., had been looking for new ways to improve the efficiency of their large and increasingly complex manufacturing operations.

Following a fact-finding trip to the US, the Lyons managers were pointed to Maurice Wilkes and EDSAC in Cambridge. One thing led to another, and soon EDSAC spawned the creation of

* Electronic Delay Storage Automatic Calculator. There are remnants of it sitting in the lobby of the University of Cambridge Department of Computer Science and Technology.

the 'Lyons Electronic Office', or 'LEO', one of the first business-focused electronic computer systems.[21] The firm's managers used LEO to help analyse large volumes of data on sales, supplies and operations.

Despite the fact that LEOs were huge (covering an area roughly the size of a tennis court), expensive (each one cost £3 million in today's money) and required a dedicated team of highly skilled technicians to persuade them to keep working, J. Lyons and Co. had demonstrated something really important by utilising them. Lyons had acted as one of those vital early adopters, taking on the risk of being one of the first to use a new technology in action, thereby showing its potential.* This demonstration led to investment in improving the technologies, which increased performance, which made the technologies more attractive to more customers, which put more computers into the market, which showed more companies their commercial potential . . . and so on.

Meanwhile, in parallel with the growth of the market for these giant computers, some very smart people were working on new technologies that would help push computers from dedicated air-conditioned rooms and onto factory shop floors and office desks.

Up until the 1950s, the most fundamental aspect of all digital electronic computers – the storing of and switching between 1s and 0s to allow logic operations to be implemented – was taken care of by vacuum tubes. These looked a bit like old-fashioned light bulbs and could control the flow of electric current between electrodes in a vacuum encased by glass, and while they worked well as digital switches they suffered from a few drawbacks: they consumed lots of power, generated a lot of heat, and frequently failed.[22] Given

* In the US, the government took on this role of early customer for the big computers manufactured by IBM – something that helped IBM become dominant in this market. In the UK, the government of the time was a bit confused as to why a cake company was making computers.

that thousands of these fragile and temperamental devices were needed to do anything useful, it's understandable why early computers were enormous and required such intensive and specialist care from people in lab coats.

Then, a couple of extraordinary technological developments changed the whole world of computing.

Things get smaller: from valves to transistors

In 1956, three US scientists won a Nobel Prize for their research on materials that sat halfway between conductors and insulators: *semi-conductors*.* The development of semiconducting materials allowed devices to be made that performed the same function as a vacuum tube – an electronic switch that could represent either a digital 1 or 0 – but which were far smaller and much more reliable. As these miniature electronic switches could be controlled to flip between both *trans*ferring electrons and acting as res*istors* to the flow of electrons, they were known as *transistors*.

The development of semiconducting materials and their application as transistors to replace the vacuum tube was a very big deal. It enabled a step-change increase in reliability and reduction in size and cost of computing technology. In combination with the development of (comparatively) user-friendly programming languages, the market for business computers started to grow. These computers were still pretty big and expensive, but their processing speed and reliability now made them a commercially viable purchase for more than just the richest of firms. The benefits demonstrated years before by Lyons using LEO were now available to many more firms.

Thanks to the process of incremental innovation, so-called

* William Bradford Shockley, John Bardeen and Walter Houser Brattain.

'minicomputers' began to appear on the market,* allowing manufacturing firms on even more modest budgets to benefit from serious computing power.

With the pace of development becoming quicker, another breakthrough was about to trigger near exponential increases in computing performance.

Things get *much* smaller . . . and integrated

The early transistors looked like a tiny cup, a couple or so millimetres wide, with three wires sticking out. Being much smaller than vacuum tubes, they enabled a whole new range of smaller electronic products – not just computers but also things like the eponymous transistor radio. But engineers managed to push things even further: they found a way to make transistors that were flat ('planar' transistors) by etching them onto a wafer of silicon. And then they got really carried away and wondered if they could make other electronic components the same way and combine them into one device. To their surprise, they found they could.

They had made an *integrated circuit.*

They then did something amazing: they managed to integrate all the circuits needed to make a working electronic computer – not just transistors, but also resistors, diodes, capacitors and much more – onto one silicon 'chip'. They had made a *microprocessor.*

The key things were now in place to launch a new computing revolution.

Except, at first, they didn't.

Strange as it might seem now, once those clever engineers found a way to miniaturise a whole computer onto a tiny chip of silicon, it wasn't obvious what people would actually want to do with them.

* Not that mini by modern standards as the early ones were still the size of a domestic fridge-freezer. But they were more mini than a room-sized mainframe.

Most computer firms who had grown big and immensely wealthy by selling large computers – and all the lucrative associated services needed to make them useful – didn't immediately see the value of miniaturised computers, and those who did were worried that these small machines could start to compete with their existing products. And potential customers were wary. What exactly would you do with one of these small computers?

Two things were needed to kick-start this new era of computing that would shortly transform the world of manufacturing.

What was going to be the 'killer app' for small computers?

There had to be some compelling use for the smaller computers that would make people think, *I must have one!* If there was no clear need for small computers, people wouldn't demand them, and if there was no demand, no manufacturer was going to set up factories to make them.

It turned out that the killer app for these small *personal computers* (PCs) was a software application that we nowadays take for granted: the spreadsheet.

The development of the spreadsheet for use on PCs did something incredibly powerful: it democratised data processing and analysis, giving even small firms using relatively low-cost PCs access to capabilities previously only available to large firms with their big, expensive systems. This led to increases in productivity, as many laborious paper-based tasks in sectors such as accountancy could now be done much more speedily using spreadsheets. New business opportunities could be more rapidly identified and targeted thanks to data analysis done at the desktop. And as costs of PCs fell, these benefits became available to even the smallest of firms.

But the distribution of computing power to the masses wasn't going to be realised unless someone started making and selling millions of those small low-cost computers. What was needed were some entrepreneurs and investors to come along and create one of Joseph Schumpeter's gales of creative destruction.

Creative destruction does it again

The scale of the opportunity that those with a bit of vision could foresee led some to pile in to fund and launch new companies, filling the gap left by the inertia-ridden big companies.* Such was the speed and scale at which markets for these PCs grew that – following the same pattern we are seeing with electric vehicles – big computer companies quickly realised they were missing a trick.

The mighty IBM corporation sprinted to launch its own PC to provide more conservative businesses with access to the benefits of small desktop computers but with the reassurance of the IBM brand,† rather than machines and software provided by those fly-by-night, wacky start-ups with names like Apple and Microsoft.[23]

Through the 1970s, 80s and 90s, the market for small computers grew from tens of thousands to hundreds of millions,[24] thanks to the coupling of hardware and software markets. New software offered exciting features that worked best on higher-performing PCs. Manufacturers in turn developed faster microprocessors and bigger memory chips and hard disk drives, hurriedly bundling these into new PCs. This encouraged software developers to add more functionality and features, which pushed the PCs to the limits of their

* These entrepreneurs included Bill Gates and Paul Allen (Microsoft), Steve Jobs and Steve Wozniak (Apple) and Michael Dell (Dell), in the US, and Hermann Hauser, Andy Hopper and Chris Curry (Acorn Computers) and Clive Sinclair (Sinclair Research) in the UK . . . along with hundreds of other founders of start-ups around the world.

† 'Nobody ever got fired for buying IBM' used to be a common refrain in the world of business computing.

speed and memory, which led to manufacturers needing to design and make PCs with even higher levels of performance, which gave software developers more scope to add more functionality, which . . . you get the idea.

It was at this point I realised I didn't have a feel for what it was actually like for those working in manufacturing while all these phenomenal changes were happening. I decided to call up two recently retired colleagues, Tom and John, who started working in factories back in the 1970s in the midst of this revolution.

From 'just too late' via 'under the radar' to democratised data analysis

In large, complex manufacturing firms, the use of mainframe and mini computers had become quite common in the 1970s and 80s. But in such firms, computers were seen as a very specialised service to which only programmers should have direct access. If engineers wanted some data analysed, they had to fill in a request form, go through an authorisation process and patiently wait days, sometimes weeks, for the results to be returned. As John put it, 'what information you did receive was limited in usefulness and always "just too late"'.

As PCs started appearing on the market, some managers saw how they could use the new-fangled spreadsheets to do analysis swiftly and right at their own desks. Often the first use of these small computers in manufacturing took place 'under the radar' by enthusiastic, younger employees like Tom and John. Not for them the patient waiting for data to be processed, nor the need for the status symbol of access to secretarial support to type up documents and prepare reports.

However, not everyone was happy with this democratisation of computing. Tom recalls the fretting by senior managers over the creation of multiple sources of data and analysis, and the confusion – along with that perceived loss of power – that such a distributed approach to computing might cause to the smooth running of manufacturing operations.

But the value of the PC as a means to improve the performance of manufacturing soon became clear. The activities of making, moving and consuming were all before long in some way enhanced through use of these smaller computers. Managers were now better able to deal with the complexity of running things inside the factory and along their supply chains; they were now able to process data quickly, and have it presented in ways that helped them spot trends and respond to changes in customer demands. We soon had an acronym soup of ways in which computers were able to support almost all aspects of manufacturing, be they CAD, CAE, CIM, CAM, CAPP or CAQ.*

But Tom and John highlighted something else very important happening at that time. In manufacturing, the arrival of smaller computers was only one part of a wider technological transformation with direct links back to Joseph Jacquard and his looms. Another part of this third revolution was about the changing distribution of tasks between people and machines.

It was about automation.

Automation is not just about robots

Back in the 1950s, US businessman and automation guru John Diebold stated, 'I think it fair to say that automation offers as great a challenge and reward as anything that industry has ever known.'[25]

That seemed rather a bold statement, and one that I felt needed some investigation. I was scratching my head working out where

* Computer Aided Design, Computer Aided Engineering, Computer Integrated Manufacturing, Computer Aided Manufacturing, Computer Aided Process Planning, Computer Aided Quality Assurance.

to start when, with excellent timing, I saw a talk advertised that was entitled 'Should We Automate?' I reckoned that if the speaker – my colleague Duncan – was going to try to answer that question, he'd first have to explain what automation was.

Looking back at the scrawls from that evening in my battered little Moleskine notebook, here's a two-point summary of what I learned:[26]

> All manufacturing processes loop through four steps – decide, act, operate, sense – any or all of which can be done by humans or automated by using some form of machine.
>
> Just because we can automate something doesn't mean that we always should.

Duncan's talk explained that John Diebold was referring to the impact of using machines for each of the steps of that *decide-act-operate-sense* loop. And Diebold was right about the scale of both the challenge and the reward. Manufacturing firms that successfully implement automation can see dramatic gains in productivity, quality and safety, coupled with reductions in cost, but actually realising these benefits is notoriously hard.

And it certainly doesn't work by simply buying and plugging in a few industrial robots.

In fact, industrial robots are just one type of device that can do the 'act' part – completing some physical 'operate' task. There are hundreds of other types of industrial automation devices, such as conveyors, motors, switches and pumps – they just lack the camera-friendly skills of a big, brightly painted industrial robot sparkily welding a car together.

To 'sense' whether the right part of an operation has been done in the right way, devices have been developed that automate touch (so that a robot can pick up delicate things without crushing or dropping them), vision (from simple barcode scanners to full vision systems), radio waves (things like radio frequency identification, or RFID, tags), audio (for example, to identify the sounds of a machine not running as it should), and smell and taste (via cleverly calibrated chemical sensors).

Finally, there are the technologies needed to decide (or support humans to decide) what needs doing and how to control the whole process. For this, one key technology is a specially ruggedised type of computer called a Programmable Logic Controller (PLC), first developed in the late 1960s.

However, all these wonderfully clever technologies created another problem: How do you integrate this plethora of automation technologies across the whole business? New software technologies appeared – along with armies of consultants that knew an opportunity when they saw one – that promised firms the nirvana of 'the one solution' to help manufacturing managers like Tom and John to join things up both within their factories and all along their supply chains.*

But something was about to arrive that would push that joined-up-edness to a whole new level.

Things start to get connected

One of my favourite TV shows when growing up was the BBC's *Tomorrow's World*. Presenters would excitedly describe and demonstrate – with varying degrees of success – some of the amazing things that experts believed would one day become part of our lives. Trawling through this programme's rich online archives a

* With some of the popular ones being CIM, MRP, MRP II, ERP, MES (Computer Integrated Manufacturing, Material Requirements Planning, Manufacturing Resource Planning, Enterprise Resource Planning, Manufacturing Execution Systems).

few years ago, I found a clip from 1967 entitled 'Home Computer Terminal'.[27] This three-minute black-and-white segment takes the viewer through a day in the life of one Rex Malik, industrial consultant, who had installed something that was, at the time, certainly not a typical household device.

I'd recommend watching this clip for two reasons.* Firstly, it is remarkably prescient in identifying many things we now take for granted in our lives. Secondly, if you put yourself in the eyes of a viewer in 1967, from a time when computers were still those giant machines locked away in air-conditioned rooms operated by people in lab coats, it looks absolutely nuts.

Relaxing awkwardly in his pyjamas in bed under the glare of the BBC film crew's lights, Rex turns from his newspaper in response to the loud clattering of a 'teletype' terminal – something resembling a giant electric typewriter on wheels – while the soothingly posh BBC voiceover tells us that Rex

> feels the business world's pulse from his bedside. Stock prices and market trends are available to him through Europe's first home computer terminal. This terminal is linked to a giant brain ten miles away in the heart of London. It's one of two Malik has installed for experimental purposes, because he wants to know if they can run his life, and his home, for him.

The voiceover explains how experts are predicting that

> in twenty years' time, all new houses will be built with special computer points, and that terminals will be cheaper to rent than today's telephones. There's no complicated language to master before he can understand what the computer is saying [. . .]

* There's a link to the clip on this book's website, www.timminshall.com.

Every day, the computer sends Rex Malik a daily reminder of where he should be, for he's stored his day-to-day diary with the brain.

With over fifty years of hindsight, that clip gave some remarkably good hints at what was coming. The story of how we actually ended up with the 'giant brain' of today's online world is very well told elsewhere,[28] but it can be summarised in three points:

In the late 1960s, the US Advanced Research Projects Agency (ARPA) creates ARPANET to connect large research computers across the country.

In the meantime, other countries had been developing their own computer networks. Thanks to the development of some clever communications protocols, these disparate networks could now be combined into one massive interconnected network – the internet.

Agreement was reached by the mid-1990s to allow commercial use of the internet, and with the development of the World Wide Web, access and usage exploded.

Those three steps (plus a few thousand other developments I have glossed over) enabled Rex Malik's 1967 demonstration – and so much more than the BBC researchers could have ever imagined – to become a reality.

There is, of course, something that even Mr Malik and the BBC didn't foresee: we now take it for granted that the 'giant brain' is accessible from the palm of our hands, almost no matter where we are in the world.

Communication goes mobile

Once you start to look, you see them everywhere. Mobile phone masts have become a common feature of our urban and rural landscapes. I am now going to show great restraint in not opening another hydrant and drenching you with facts about the (well, I think) fascinating world of macro-, micro-, pico- and femto-cell towers, plus a few thousand low-orbit satellites,[29] whose integrated operations allow us to binge-watch our favourite shows, make FaceTime calls, or indeed monitor the operations of a factory, wherever we may be.[30]

Two numbers give a snapshot of the scale of the growth of the market for mobile phones over the past four decades:

0 = Number of mobile phones sold in 1982.*

1,300,000,000 = Number of mobile phones sold in 2023.[31]

Since the 1980s, we've gone from no mobile phone accounts to (on average) one for every person on the planet.[32] And, of course, after the arrival of the smartphone around 2007, almost all of these devices moved beyond mere voice and text: they were now mobile internet-access machines. Heading into the 2010s, the world had around one billion PCs, five billion mobile phone users and an internet that kept growing, morphing and becoming ever more connected to almost every aspect of our lives.

By the mid-2010s, the scale and impact of this relentlessly expanding connectivity on how we make, move and consume led some to propose that we were at the start of yet another industrial revolution.

* There had been mobile, radio phones available for decades, but the first commercially available hand-held cellular phone was the Motorola DynaTAC, and this was launched in 1983.

So, is this Fourth Industrial Revolution really a thing?

Writing these words in 2024, it's still hard to see whether we really are in the early stages of a new revolution, or just an accelerated evolutionary drift into a highly caffeinated version of the third one. I think it depends on how we define things. Here's a description of the 'Fourth Industrial Revolution' from the man who popularised the idea, Professor Klaus Schwab, founder of the World Economic Forum:*

> Previous industrial revolutions liberated humankind from animal power, made mass production possible and brought digital capabilities to billions of people. This Fourth Industrial Revolution is, however, fundamentally different. It is characterized by a range of new technologies that are fusing the physical, digital and biological worlds, impacting all disciplines, economies and industries, and even challenging ideas about what it means to be human.[33]

Well, that all sounds very exciting, doesn't it? But what does it mean for manufacturers today? There seem to be two important features of this new manufacturing context that really are happening now and having an immediate impact on our lives.

Even compared to a few years ago, we have orders of magnitude more data from right across the whole make-move-consume spectrum. And with all that data – coupled with remote access to staggeringly fast low-cost computing power – manufacturers can do previously undreamed-of things to address our wishes and wants.

First, we need to look at how manufacturers are getting access to all this data.

* The organisation that assembles some of the world's elite to a conference Davos each year.

Connecting everything to the internet

Kevin Ashton, then a brand manager at consumer products manufacturer P&G, made this statement in 1999: 'We need to empower computers with their own means of gathering information, so they can see, hear and smell the world for themselves, in all its random glory.'[34] What he was talking about was the need for an 'Internet of Things'.

The Internet of Things (IoT) is real, and describes exactly what it is. If the internet is about connecting computers, the IoT is about putting technology into objects so that they can generate, transmit, receive and do things with data. These 'things' could be your phone, watch, car, burglar alarm, baby monitor or fridge.*

IoT technologies – tiny transmitters, receivers, sensors, processors – are now ubiquitous across the manufacturing world. There, rather predictably, they are known as Industrial IoT (IIoT) devices. IIoT technologies are what allow data from those jet engines roaring in the frigid air ten thousand metres above the Russian tundra to be captured and transmitted to be analysed on screens in an airconditioned room back in the UK. They're also what allowed Mo, the delivery rider, to know exactly when and where to pick up and deliver my order, and me to get live updates on my phone of the whole process. And IIoT technologies enable data from all those robots and AGVs in the Zeekr factory in Ningbo to be co-ordinated to operate in the most efficient way possible as they attempt to respond in a flash to changing customer demands or a problem in their supply network.

In this hyperconnected world, the scale of almost instantly accessible data coupled with vast computing power means that we

* For some reason, the 'internet enabled fridge' is often given as an example to explain the potential of IoT. I have no idea why.

are starting to see the emergence of something amazing. We can create parallel universes.

Creating 'digital twins'

Well, 'parallel universes' is perhaps overselling things a tad. But we are able to do some pretty impressive things. Thanks to all the IIoT devices attached to those jet engines, for example, coupled with ready access to supercomputer-like processing power, engineers can do more than just monitor what is going on mid-flight. They can create a completely virtual version of each engine, a *digital twin*.

With this fully operational virtual replica of the real engine, they can fiddle with various parameters in a low-stakes environment. Perhaps one of the real physical engines is burning a bit more fuel than expected. The engineers can scratch their heads and try out some possible solutions on the digital twin. If one seems to solve the problem, they can then get ready to do the same on the real engine when the plane lands. Unsurprisingly, the idea of the digital twin is attracting a lot of interest from big manufacturing firms across many sectors.

So does this ability to rapidly capture and analyse data on this massive scale justify the use of the 'new revolution' tag? Perhaps, but whether it is a revolution or not, there are four big issues we need to consider very carefully as we head further into this new digitally hyperconnected era for manufacturing.

1. The future has arrived – it's just not evenly distributed yet[35]

A couple of years ago, I was sitting with some students in a cramped meeting room off to one side of a small factory. The company made sensors for detecting various types of pollution. The boss was explaining his challenges: the business was growing fast but facing some operational problems. As more orders came in from customers, he was finding the process of 'job tracking' – knowing how far individual customer orders had progressed through each stage of assembly – a bit hard to manage. The students, perhaps a little overexcited by a recent lecture on digitalisation, leapt in with suggestions of how IIoT technologies could be used to track components and part-finished goods through the factory, with screens on the shop floor used to make all the data easily visible so that problems could be spotted quickly, perhaps even linking in to mobile phones to alert the boss when things weren't flowing as they should. The manager listened politely, cleared his throat and pointed out that, while that all sounded great, most of his data was currently on various bits of paper, so maybe the most useful thing they could do would be to set up a spreadsheet and type numbers in for him.

The students' faces blended crestfallen with sheepish. They had been confronted with the reality of something often overlooked with the arrival of this new era. Most manufacturing firms are not giant, well-resourced companies like Rolls-Royce or Volkswagen or AstraZeneca, nor are they super-rich F1 teams or well-invested high-tech start-ups. Some 99.9 per cent of manufacturing firms are classified as 'small or medium-sized enterprises' (SMEs) – small in scale, skilled in a specific area, but with limited resources to try new things.[36]

That's quite a problem. This latest industrial revolution is going to be off to a slow start if hundreds of thousands of smaller manufacturing firms don't have access to the required technologies or skills to use them.

2. Can I give it a kick?

This was the rather alarming question put to the nice man demonstrating an amazing bit of robotic technology at a trade show.* The technology in question was 'Spot', a robotic dog you can buy now from US company Boston Dynamics. Spot was being showcased as an example of a robotic technology that could be sent to do dangerous or boring things that humans wouldn't want to do, like monitoring facilities in remote or dangerous locations.

These new technologies offer loads of potential benefits for us as consumers as well as those of us that work in this new data-drenched manufacturing world.[37] But just as some firms will be able to thrive in this world while others struggle, so too will individuals.

A good example of this can be seen in one aspect of digitally enabled manufacturing logistics: the $150 billion per annum business of home delivery services.[38]

Turancan Salur, the UK general manager of Getir (a home delivery service) said in 2021 that his type of business was 'democratising the right to laziness'.[39] Why should only the wealthy have the right not to go to the shops? If this new industrial revolution enables more equitable access to home delivery, that's all for the good, right?

But there's another angle to this.

The seamless digital experience of getting any food delivered still relies on some very people-based activities. Mo, the pizza delivery driver from earlier, needs to spend many hours on his scooter and weave his way to and fro along the dark, rain-drenched, pot-holed streets of Cambridge to deliver food to people like me in my warm home, in return for modest payment. This part of the manufacturing world is 'a cost-intensive business that is low-margin and scale driven'.[40] Meaning? It must be big – hence firms spending whopping amounts on advertising to leverage the appeal of Robert De Niro

* In case you were wondering the answer was '*No!*'. And the person asking the question wasn't a psychopath: he wanted to know how well Spot could respond to being pushed off course. And if you wish to know the answer, I've put a video about this on the book's website, www.timminshall.com.

and Snoop Dogg – and it must be super efficient, minimising costs wherever possible if it is to generate any profit. To achieve that, this whole sector relies on hundreds of thousands of people worldwide like Mo willing to hurtle around city streets in all weathers.

3. Automating our decision making

Back in November 2022, Carl, one of my research students, offered to give a demonstration of the then newly released ChatGPT during one of our research group meetings. At first, it just seemed amusing. We did the nerdy thing of getting it to write messages in the styles of different people. As a 'large language model' (LLM), ChatGPT uses AI to generate human-like responses by learning patterns from a huge body of stored text. These LLMs offer benefits such as more efficient customer services (a chatbot that responds in a human-like manner to queries)[41] and improved efficiency of many white-collar jobs.[42] They have also led academics to the alarming realisation that students can now complete some assignments in seconds rather than hours.

And AI is certainly having an impact on manufacturing. Back in Chapter 1, you'll recall how my colleague Alexandra provided eye-wateringly complex images of the reality of manufacturing supply networks. She and her team have in fact done much more than just visualising the links between all the companies involved in making, say, a car or an aeroplane. These supply networks are so fantastically complicated that no single company can be completely sure of what is going on, nor can any manufacturing manager be sure how these networks will react to disruptions like a hurricane in Mexico, geopolitical tensions in the Far East or a global pandemic. By using AI, however, Alexandra and her team can reveal so-called 'hidden dependencies' from analysis of mind-bogglingly large volumes of data from a wide range of sources and help managers prepare for and respond to disruptions.

In this way, AI is being used to help people working in manufacturing firms deal with the 'decide' part of that four-step loop described earlier from Duncan's talk.

And that's where things could get a bit concerning. As we've seen, automation has been a feature of manufacturing since the First Industrial Revolution – Jacquard's loom was a form of robotic automation, one that used a machine to replace a skilled worker. What is different now is that the machines have some form of intelligence and can play an increasing – and sometimes poorly understood* – role in decisions made in ever more complex manufacturing systems.[43]

If we try to imagine a future where smaller firms can fully participate in this brave new manufacturing world, where a fair balance can be struck between people and machines, and where AI is used to augment rather than replace our ability to make decisions, we have to assume that the digital technologies and systems that underpin this new industrial revolution are going to actually work reliably and consistently.

And that is perhaps a naive assumption.

4. Two big machines – two big problems

Whether we are producers or consumers, our participation in this new industrial revolution is reliant on the flawless operation of two giant 'machines'. And, unfortunately, these machines are far from flawless.

The internet has not only impacted the world of making, moving and consuming, it is itself arguably the biggest product of the manufactured world. Though we now use soft, comforting words

* This is why the concept of Explainable AI (XAI) is so important: XAI allows human users to understand and trust the results and output created by algorithms. The opposite is Black Box AI, where decisions are made without the user being able to see how they were made.

like 'the cloud' to label – what is to most of us – the completely invisible places where all our data-encoded desires and needs get fulfilled, the internet is very much a hard, physical thing. It is a vast network of cables,[44] connections and, basically, 'other people's computers', all working in a sort of harmony without anyone really being in overall control.[45]

The internet had to be built, has to be maintained and is in a constant state of development and improvement. It has delivered extraordinary benefits to almost every aspect of our lives, but has done so at a significant cost. Hundreds of billions of dollars need to be invested every year to ensure that the performance of the network keeps up with our insatiable demands for greater, faster, more secure connectivity.[46] Enormous amounts of energy and water are consumed to stop it overheating and to keep it running.[47]

The internet was designed to have resilience at its heart,[48] but it has a number of vulnerabilities that can be exploited, and such exploitation can take the form of me being scammed when trying to buy some new shoes online (somewhat embarrassing) to ransomware attacks that lock clinical staff out of thousands of computers in hospitals (definitely expensive and potentially life-threatening).[49]

Our manufactured world, now and in the future, depends on the reliable operation of the behemoth that is the internet. And that requires ubiquitous connectivity and the seamless, secure and lightning-fast flow of petabytes of data between the smallest of the billions of IIoT devices and the vast computation and storage network that is the cloud.

The second huge 'machine' to which so many aspects of our lives have become inexorably linked is the immense and globally distributed network of highly specialist firms that make up the global semiconductor industry. Without this second machine to feed the first, we'd have no modern world. And I chose the word 'feed' intentionally. The thousands of servers that power the cloud and all the switches that steer zettabytes* of data are in a constant process of upgrading. They all need powerful microprocessors to allow them to keep pace with our demands, from high-definition video streaming to our phones, to microsecond-fast decisions in financial markets, zero-latency remote control of robotic surgery and powering AI.

The semiconductor industry is simply extraordinary in what it has achieved. Earlier in this chapter we talked of those smart engineers who managed to etch thousands of transistors onto a tiny wafer of silicon, and thus make smaller, personal computers possible. One of the first proper 'integrated circuits' that formed a microprocessor – the Intel 4004 – contained an impressive 2,250 transistors. Computing power that had previously occupied a whole room or at least something the size of a big fridge was now smaller than the nail on your little finger. Fast forward from the 1970s to 2024 and the numbers are . . . well, I had to ask my researcher Lizzy to check this twice to make sure it was correct.

Ninety-two billion.[50]

That's the number of transistors on the M3 Max microprocessor used in the 2024 version of Apple's Mac desktop computer.[51] The actual chip itself is still the size of your small fingernail. How many technical challenges had to be overcome to squeeze so many components down to such a ludicrously small scale? What an achievement. This is engineering at an unbelievable level of precision and complexity.

But this isn't about doing this remarkable thing once in a lab. Manufacturers must be able to repeatedly – and cost-effectively – make perfect versions of billions of chips. To do that requires not only the most highly talented engineers, but also the most remarkable of machine tools, like the TWINSCAN NXE EUV machines produced by Dutch company ASML.

* 1,000,000,000,000,000,000,000 bytes. That's equivalent to the data stored on a billion 1TB computer hard drives.

The cost, scale and complexity of these machines were eloquently described by tech journalist Will Knight as 'kind of bonkers'. For example, each of these machines is the size of a bus, costs $150 million and has 100,000 parts (including two kilometres of cabling), and moving all the parts needed to install one of them at a customer's production site requires forty shipping containers.[52]

This whole industry is, like those ASML machines, huge, complex and kind of bonkers. And these three terms provide useful labels to help me summarise this industry for you.

Huge

The following numbers give a sense of scale of the industry. Semiconductors are now the world's third-most traded product behind oil (crude and refined) and cars.[53] Annual worldwide sales of semiconductors are $600 billion.[54] The cost of developing a single new chip is at least $1 billion.[55]

A single semiconductor factory costs $12 billion and the major companies – like Intel and Samsung – each run multiple factories scattered around the world.[56]

Over the next ten years, experts predict that the industry will need to invest about $3 trillion – about the same as the annual GDP of France – to meet our insatiable demand for more and better semiconductors.[57]

Complex

At the top level, the semiconductor industry is quite simple. It is just about designing, making and testing. To design a new chip, you need some very expensive specialist software called 'electronic design automation' (EDA). Then you go on to the actual making, a process which starts in a mine and moves to the casting of silicon ingots, which are sliced into thin wafers and have the electronic circuits 'printed' onto them, before assembly, packaging, final testing and shipping.[58]

These deceptively simple 'major' steps mask frankly staggering levels of complexity. Depending on the actual chip, there are between one thousand and two thousand sub-steps in the overall manufacturing process.[59] From the start of the process to the end requires over fifty different types of highly specialised, extremely expensive machines (like those monster ones from ASML).

Due to the scale of costs, constant advances in technology and ever evolving markets, the semiconductor manufacturing industry has fragmented. While there are still the giants like Intel, Samsung, Infineon and Texas Instruments who do everything, many other firms have chosen to become specialist providers of one or more parts of the semiconductor manufacturing design-make-assemble-test processes. There are now hundreds of companies – nearly five hundred major ones at a recent count[60] – knitted together to form a worldwide 'semiconductor ecosystem'.

Just take a moment to look at your phone. It is likely that the tiny wafers of silicon sitting at the heart of that phone came from a quartz mine in the Blue Ridge Mountains of North Carolina. Why there? Because that's the location of some of the purest quartz, from which you can extract some of the purest silicon in the world. It is also very likely that the layout of the billions of transistors on that chip uses a design from UK-based ARM, that at least one of the many chips that make your phone work was produced somewhere in a giant city-like industrial park in Taiwan, and that at least one of them was made there in a foundry run by local company TSMC using machine tools made by that Dutch company ASML.

These chips are a vital input to the supply chains not just for consumer electronics and computers, but for thousands of other products in dozens of other sectors. This super-complex semiconductor ecosystem has become the lifeblood of so many industries – from cars to medical devices, aircraft to farm machinery. If it wobbles, the impact on manufacturing across the globe can be substantial and far-reaching.

Bonkers

A news headline in mid-2023 read: 'Why Nvidia is suddenly one of the most valuable companies in the world'.[61] The reason? Nvidia is a semiconductor company whose success, until recently, was based on its skill at designing graphics processing units (GPUs), chips that do the super-quick parallel processing needed to make video games realistic.

As demand for processing power increased with the use of AI-based tools (of which ChatGPT is just one recent example), Nvidia developed tools to allow its GPUs to be programmed for AI. This proved to be a really clever move: it allowed them to dominate the market for AI GPUs (they had 95 per cent market share in 2023) and join that elite club of firms valued at over $1 trillion.[62]

But now the loop is closing. The use of all AI tools requires vast and blisteringly fast processing power. That power requires the use of supercharged chips like those produced by Nvidia, but to design ever more powerful chips to cope with the demands of AI, semiconductor companies themselves need to use AI to accelerate their chip design and manufacturing activities.[63] Companies like Nvidia need the chips that power AI to be more powerful to allow them to use AI to design and make more powerful chips for AI, the use of which pushes demand for yet more powerful chips, which require . . . At this point you might, like me, find that your head is starting to hurt and you need a little lie-down.

Assuming you are back from your break, I'm going to now draw out the main points from this surface-skimming exploration of this nascent Fourth Industrial Revolution and why it will affect so many aspects of all our lives. It does seem to be a real thing, and it has the potential to deliver manifold benefits to those who make and ship things and those who buy and use those things. However, there are a few problems that still need ironing out.

Many smaller manufacturing firms will struggle to engage in this brave new world unless they are somehow helped to adopt the new technologies. And while digitally enabled automation is helping remove humans from dirty, dangerous and dull jobs, as things currently stand, this new revolution still seems to need an awful lot of people doing low-reward jobs. AI is playing an increasingly pervasive role in many aspects of this new manufacturing era, from the prediction of machine failures, to the optimisation of local (pizza) or global (car components) supply chains, to the ability to accurately 'chase demand'.

This new manufacturing world requires the consistently reliable operation of two staggeringly complex and constantly evolving 'machines', the internet and the global semiconductor industry, both of which have many known flaws . . . and probably a whole load more that we will soon discover.

As I was driving along a residential street in Cambridge, something on the pavement caught my eye. A group of adults and small children had circled around to admire what I thought was a baby in a buggy. As I got closer, I realised it was something quite different.

They were unloading a small autonomous home-delivery robot that had brought goodies from the local grocery store. These are being trialled all around the world and have the potential to take on 'last mile' deliveries.[64] These AI-driven robots could offer an alternative to low-paid delivery riders having to work long hours to enable the ongoing 'democratisation of laziness'.

It's not much of a stretch to imagine a world where IoT devices in our fridge feed data into the internet – the manifestation of Rex Malik's 'giant brain' – which then uses AI to find the optimal source for my depleted groceries and automatically sends out a robot to arrive at my front door.

Is that a vision of utopia or a step too close towards Skynet-like loss of control?

My wife pointed out that there is probably still a bit of time before these machines take over. A few days after my sighting of the grocery delivery robot on the street, she also spotted one of them in rather a different situation. The little robot was stuck trying to cross a busy road. It was waiting for a human to come along and press the button at the pedestrian crossing.

The machines still need our help. For now.

So what will the role of us humans be in this hyperconnected new version of manufacturing? To understand that, a good place to start is by looking at how the manufacturing world supports us when we need it most.

7: Merge

How the boundaries between making and consuming are becoming ever blurrier

Up until 2020, my engagement with the world of healthcare had been mercifully superficial. Most interactions arose from my children's relentless enthusiasm for crashing into hard objects or catching some new bug. But as my sticky fingers syringed Calpol into a sobbing mouth, or my teeth tore open a plaster to staunch a bleeding limb, I did wonder why things clearly needed in a hurry were so fiddly to use. When I popped yet another paracetamol tablet to numb a random ache, I might have idly wondered how much of this particular drug was produced globally each year,* and lying nervously inside a humming body-scanning machine, I couldn't help but puzzle how they got this thing through the door. Other than that, like most people, I really didn't give much thought to medicines, medical devices and other healthcare supplies. When they were needed, I was always very happy to find that they were just there.

In 2020, this changed.

The COVID-19 pandemic abruptly demonstrated the critical but complex role of manufacturing in keeping us alive and healthy. The sudden absence – or failure – of everyday healthcare supplies and systems directly affected us all. Without access to simple protective equipment, healthcare workers were forced to do their jobs at great personal risk. Without vaccines, we couldn't protect ourselves and our families to allow normal life to resume.

* In case you are interested, it's around 160,000 tonnes.

With hospitals suddenly overwhelmed, we simply didn't have operational healthcare systems.

But while we despaired at the rapid disappearance of even the most basic of supplies – how could we not have enough medical gloves? – we also rejoiced at heroic tales of communities coming together to make and donate supplies to their local hospitals and care homes. Our newsfeeds provided us with the real-time drama of the global race to develop effective COVID-19 vaccines. And we saw temporary hospitals being built at a blurringly fast pace to cope with the anticipated tsunami of incoming patients.

These activities – managing medical supply chains, developing medicines, building hospitals – are actually all quite routine things. Every day, new medicines are developed, made and shipped around the world. The manufacturing of medical supplies and devices is a massive and growing industry,[1] and the building and operating of hospitals is one of the largest areas of spend in most countries.[2] These are regular, day-to-day activities that are planned and slowly, methodically implemented, albeit mostly behind closed doors.

COVID-19 smashed open those doors, marched in and jammed its thumb on the fast forward button.

The truth is that numerous cracks in our healthcare systems had been appearing years before the pandemic. COVID-19 just widened those cracks and revealed to us all how alarmingly fragile these systems are. However, there was a positive side to this otherwise tragic and traumatic story. The pandemic also showed that things could be done differently in the healthcare parts of the manufacturing world. This crisis accelerated the realisation of exciting ideas that had been lurking in corners for years, and it also took us back in time, putting the 'local' back into manufacturing as we were forced to shorten the distances between making and consuming.

So how did that happen, and what might this tell us about how our manufacturing systems could be changed for the better?

How a crisis revealed something about the future of manufacturing

Once the scale of the impact of COVID-19 became obvious, our institute's staff and students – like so many other people around the world – threw themselves into doing whatever was needed to support our local healthcare system.

Throughout the crisis, I tried to document what we and others were doing and to archive news stories and data that captured the reality of how the situation was evolving in real time.* Once the worst of the storm had passed, I sat down to review the notes I'd scribbled in my battered little notebooks, coupled with the stories, messages and data I'd kept. As I trawled through and reflected on all this curated information in the context of everything we've covered so far in this book, something rather surprising started to form in my mind.

That crisis had given us a glimpse of one possible future of the manufacturing world. Within this new world, boundaries between manufacturing and other parts of our lives become blurrier, and you and I are no longer passive consumers of outputs. Individually and collectively we become increasingly integral components of our manufacturing systems.

I now want to take you back to the midst of the pandemic crisis to explore the manufacturing issues that led me to this conclusion.

Making medicines

Thousands were dying. Populations were locked down. Healthcare systems were at breaking point. Without a viable vaccine, there was no obvious way of ending this nightmare.

But as I listened to the news in early 2020, I was a bit confused.

* I've put some of the articles and case studies we produced on this book's website, www.timminshall.com.

The updates we were receiving from scientists and government officials seemed to suggest that the vaccines would be with us within a few months. That seemed ludicrously optimistic to me. I thought new medicines normally took about a decade to get from lab to patient. Was this too good to be true?

It turned out it wasn't.

By December 2020, the Oxford University/AstraZeneca vaccine was approved for use in the UK and factories were set to produce millions of doses. That whole process had taken less than a year.

How on earth had that been possible? I realised I couldn't answer that question until I had filled in some of the pretty large gaps in my knowledge of medicine development. I contacted my colleague Ettore, recommended to me as the go-to person for information on the latest developments in medicine manufacturing, who sent me away to read the equivalent of a Ladybird book of how we make medicines before agreeing to answer any of my questions.

Here's what I learned.

Why it takes so long to come up with a new medicine

Developing a medicine is usually not only very slow but also very expensive. The time from TV news featuring a white-coated, purple-gloved scientist explaining their research breakthrough to something that can be picked up in the pharmacy or jabbed into your arm is typically ten to fifteen years, and it will cost the company doing this *at least* $1 billion. The process takes this long and costs so much because three difficult questions need to be answered unambiguously:

Does this new treatment really work?

Is it safe?

Is it better than what we already have?

First, scientists will identify or create something that seems to interact in a useful way with 'targets', things within your body that cause the disease or condition.* When something promising has been found, they commence a clinical trial, which has three phases.

In Phase 1, the scientists show that what they have produced doesn't harm healthy people. In Phase 2, they show that it treats the condition in a small number of real patients. Finally, in Phase 3, they test that it is effective at treating thousands of patients. Each phase can take many months or, more typically, several years. Very, very few new medicines climb all steps from lab to patient – I was gobsmacked to learn that 96 per cent of potential treatments entering clinical trials fail to lead to a usable medicine.[3] With odds like that, the chance of developing an effective COVID-19 vaccine in super-quick time didn't seem terribly likely.

For the tiny proportion of potential medicines that successfully exit the far end of the clinical trials, the journey to the patient is far from over.

Medicine companies next have to convince government regulators that every single one of the little tablets, capsules or vials they are going to make will be identically safe and effective. They have to demonstrate how production will be scaled up from the relatively small number of doses needed for the clinical trials to the millions or billions needed to treat patients around the world. To do this, they have to show compliance with a set of rules with the rather dull title of 'Good Manufacturing Practice', or 'GMP'.[4]

GMP shows how the process for making a particular drug will be followed – in obsessive detail. Companies must define every ingredient, where they will be bought, the exact machines that will be used in the factory and where they will be positioned, even

* For COVID-19, the targets were so-called 'spike proteins'.

the clothes the workers must wear when completing each step. Then the route by which this new medicine will be distributed from factory to patients must be mapped out, showing every step involved in that particular supply chain. There are also tight rules for all these logistical activities, and these are called (you guessed it) 'Good Distribution Practice' (GDP). Pages and pages of GMP and GDP checks have to be completed and hundreds of boxes ticked to gain approval for the medicine to be used.

Then comes the final step: getting someone to actually buy it.

That 'someone' typically works for a public or private health provider or an insurance company. They will trawl through reams of evidence to decide whether this new medicine is appropriate value for money and negotiate what they consider to be a fair price.

'Value for money' and 'fair price can be particularly thorny issues.

The manufacturers (or, more precisely, their shareholders) want to recoup their billions of investment as quickly as possible, and so will want to charge the highest viable price. The insurance company or health service provider must make hard-nosed, economics-based decisions. They have to weigh up many conflicting issues. Should they buy medicines that might prolong life for a small number of patients but cost hundreds of thousands for every course of treatment? Or should they spend the money ensuring lots of vulnerable people get a more basic and cheaper treatment for long-term conditions? These choices will determine the quality of people's lives and, in some cases, who lives and who dies.[5]

Needless to say, these negotiations are neither straightforward nor quick. You can see how those ten to fifteen years just fly by.

A miracle happens

'The Oxford-AstraZeneca vaccine has been approved for use in the UK, with the first doses due to be given on Monday' – that headline from the BBC News website on 30 December 2020 summarises an absolutely phenomenal achievement.[6]

From development in a lab to approval for use in less than one year. And in just two years, over two billion doses of that vaccine were successfully administered worldwide.*

That story – and those of all the other vaccines that were subsequently approved – led to many people asking a very reasonable question. If we've just developed new vaccines in less than twelve months, why does it normally take at least ten years to get a new medicine to market?

One reason for this speed was the presence of a single, compelling, urgent goal. When you've got one of those, especially when it's as all-consuming as dealing with a global pandemic that has halted normal life for billions, the regular speedbumps seem to get magically flattened.

Doing things quickly is dependent on attitudes to risk: not risks to patient safety – these are absolutely inviolable – but financial risks. During that pandemic, governments, universities and companies took huge gambles, rapidly signing off millions to underwrite development costs for things that could – and in many cases did – fail. That is not something with which taxpayers, shareholders and research funders are normally comfortable, but with a more risk-tolerant approach to funding, the delays between each of the regulatory steps were reduced. Some stages in the process could even be started in parallel – *if we assume this stage is going to be successful, we can start work right now on the next one*. Of course, if the first assumption proves to be wrong, you have to scrap everything. If it is right, however, you are well ahead of the game.

* I'd thoroughly recommend reading this story in detail, told from those who led this project, Sarah Gilbert and Catherine Green, in their book *Vaxxers: The Inside Story of the Oxford AstraZeneca Vaccine and the Race against the Virus* (2021, Hodder & Stoughton).

This accelerated approach ultimately led to a remarkable announcement being made in the spring of 2022. While media attention was focused on Russia's invasion of Ukraine, the health-care ministers of the world's major economies met quietly in Oslo to publicly commit $1.5 billion to an extraordinarily ambitious challenge. Could we gear up our medicine manufacturing systems so that they could deliver safe and effective vaccines within one hundred days of any new epidemic or pandemic threat – a so-called 'Disease X' – being identified?[7]

The success of developing vaccines at speed during the pandemic had reframed the medicine manufacturing world's perception of what was possible.

I now had a somewhat scrappy yet pleasingly detailed sketch of the marathon-like stages of developing a new medicine. Underneath that timeline, I had drawn another that showed the super-squished version which reflected the sprint approach used for the COVID-19 vaccine. Something that normally takes the width of an A4 page to visualise had been squeezed down to two centimetres.

Impressive though that time compression clearly was, it raised another question. If the development and approval of new medicines can be so tightly compressed, are there ways of improving the next part of the process, getting the treatment from factory to patient?

Borrow someone else's supply chain

Back in 2009, the Tanzanian government was facing a problem. An audit of their medicine distribution system had revealed many bottlenecks that were preventing treatments reaching those who most needed them. Tanzania is the largest country in East Africa with a population of over 55 million, and outside the capital, Dar es Salaam, medical services were patchy, with remote health centres frequently suffering from a lack of medicines. Healthcare experts pondering ways to address this noticed something really interesting. While links in the medicine supply chains to remote locations kept breaking, the chains of delivery for other products seemed much more robust. And one in particular stood out: Coca-Cola.

That was the genesis of 'Project Last Mile'. With funding from various global organisations, a partnership was established between Coca-Cola and the Tanzanian government's Medical Stores Department. Coca-Cola's impressive skills in delivering sugary, fizzy brown water to almost every remote corner of the globe with superb efficiency helped enhance Tanzania's medicine distribution system. The result? Drug delivery times cut from thirty to five days. And it was estimated that 90 per cent fewer drugs were expiring in warehouses as they sat waiting too long for onward shipping to where they were needed.[8]

Move the factory closer to the patients

Sometimes the fragility of medicines, or the nature of the medical situation where they are needed, can put strain on even the most well-organised supply chains. Certain medicines must be kept at a very precise temperature at every step from factory to patient – if they get too hot or too cold at any point in their journey, they arrive unusable. For some COVID-19 vaccines, this was a huge issue, as they needed to be shipped and stored between −90°C and −60°C. Maintaining a strict cold chain throughout such long, complex supply chains can be very challenging. Other medicines might be needed in high quantities for a short period of time in a very specific location, like a vaccine to deal with a localised outbreak of a virus to stop it spreading. Organising supply chains to converge with high output in a specific place can also be extremely difficult.

One potential solution to both of those challenges is *distributed manufacturing*: moving the factory closer to those who need the product. This is now being done for medicines through 'modular GMP facilities': complete, quality-certified mini factories squeezed into shipping containers and freighted off to where they are needed. Once they are plugged in, they can start manufacturing fully approved medicines on the spot.

This is an idea that has been bouncing around for many years. The problem has always been getting beyond the *What a great idea!* stage through all the tedious details of actually making it work. COVID-19 – and the anticipation of future pandemics – has accelerated its development. One firm, BioNTech, has designed mini vaccine factories – or 'BioNTainers' (who names these things?) – that could be sent to places that lack their own vaccine manufacturing capabilities.[9]

As it often does, Google Autocomplete led me to something I found quite extraordinary. While searching for more information on new approaches to medicine logistics, I started to type 'Effectiveness of medicine . . .' but before I had managed to add 'distribution', the suggestions opened up a whole new dimension to my exploration of medicine manufacturing.

'Drugs on the market work, but they don't work in everybody'[10]

It turns out there's only a fifty-fifty chance that most of the medicines we take are actually doing anything useful for us.[11] Why? Because they are designed to treat the average person, but each of our bodies is a unique bundle of features influenced by our genes, lifestyle and other health conditions (or 'comorbidities').

I know what you're thinking: this seems appalling from a manufacturing efficiency standpoint. Why spend all that money developing, making and shipping things that aren't going to actually do any good for a large proportion of consumers?

This problem is widely recognised in the world of pharmaceuticals. The good news is that there are some rather extraordinary things under way to address this. The medicines manufacturing world is in the midst of a transformation, one that could result in each and every one of us being at the epicentre of our own little bespoke medicines manufacturing world.

We are speeding towards the era of *personalised medicine*.

Building a manufacturing world just for you

Some versions of personalised medicine develop treatments for a group of patients with a common set of characteristics, such as genetic traits or comorbidities. But in its purest form, personalised medicine is just that: specific treatments for individual patients at a particular time in their lives.[12]

Within this pure, individual-centric version of personalised medicine, the actual making and distribution of medicines look very, very different.

My colleague Jag and his team offer this snapshot of how manufacturing in this new world might look:

> Imagine a patient that has complex medical issues requiring multiple drugs with multiple dose variations over time. Remembering to take the correct drugs at the correct time is a problem. The local use of 'pill-printers' that would simply combine pre-formulated versions of the drugs together in either a liquid or solid form using a liquid or powder handling robot into a single dose. The 'printer' would be programmed with the prescription of the patient and the drugs mixed together in a binder matrix and then formed into the pill.[13]

This may all sound a bit far-fetched compared to where we are now, but think of it like the shift we've already experienced from broadcast news on the TV or radio to the personalised newsfeed on our phones. And think how far the world of advertising has moved from standardised, one-size-fits-all adverts to individually targeted content creepily following our every online step. Personalised medicine is the medical version of those transitions.

So, how might this future world of personalised medicine production actually look for you and me? You might yawn your way to the kitchen in the morning and, while waiting for the kettle to boil and click itself off, press 'Print' on your Nespresso-esque Pill-O-Matic machine sitting on the worktop. It whirrs and clunks, and a few seconds later out pops a pill that is just for you, containing the precise cocktail of chemicals needed to deal with your specific conditions at that very moment.

Personalised, local, immediate. What could possibly go wrong with that?

Well, quite a lot.

Back in 2016, I mentioned the Pill-O-Matic idea during a radio interview, and about a nanosecond later a listener posted online, 'Really? How long before someone hacks data on one of these machines & starts messing with things?'

And people are right to be sceptical. As we have seen, every aspect of the current world of medicine development, production and distribution is tightly controlled, located in very specialised facilities and managed by highly skilled scientists and engineers. This is all for a very good reason. The consequences of getting even one tiny part of this process wrong can be fatal for patients and financially ruinous for the companies involved. Why would anyone want to do anything that potentially weakens the safety nets provided by Good Manufacturing Practice and Good Distribution Practice?

—

As one hand fiddles with the half-used pack of paracetamol on the kitchen table, I scroll through websites that look deeper into the future of medicine manufacture. The text on one page catches my eye.

> Imagine being able to get a faster diagnosis of a condition based on your unique situation, be given personalised treatments based on what would be most effective for you and experience fewer, if any, side effects . . . or even move away from simply managing an illness once you are sick to promoting health by predicting certain conditions and preventing them developing in the first place . . . by understanding the role our DNA plays in our health, it can help us transform how we think about our healthcare . . .[14]

That extraordinary view of the future is not from some wacky, pseudo-medical website: that comes from the UK's National Health Service. This vision of the future moves us way beyond just becoming efficient at making more effective medicines. We would be better at predicting and preventing disease, more precise and timely with our diagnoses and better placed to undertake more targeted interventions. There would also be a bigger role for us as patients, opening up a whole smorgasbord of extraordinarily complex ethical issues from ownership and (mis)use of personal data to equity of access to treatments.[15]

It is a future vision where we, as patients, are becoming increasingly merged into the medicines manufacturing world.

Colleagues I spoke to working in medicines manufacturing are convinced that some version of these ideas is plausible, but it isn't likely to happen any time soon. Reading from my notebook, I see that I reported a colleague's wonderfully understated view that 'there are one or two issues still to be dealt with'.

The examples just discussed have shown how the development

of one type of medicine was reimagined during COVID-19. We have also seen how there are broader transformations under way that are changing the very nature of how we develop and deliver treatments for us as patients.

Are we seeing changes of similar scale in other parts of the healthcare manufacturing world?

Looking at how we were forced to change the way we made medical devices during the pandemic is very revealing. The stories I am about to share with you could be harbingers of how the manufacturing world will evolve into some version of the hype-laden Fourth Industrial Revolution.

Making medical devices

> The World Health Organization has warned that severe and mounting disruption to the global supply of personal protective equipment (PPE) – caused by rising demand, panic buying, hoarding and misuse – is putting lives at risk from the new coronavirus and other infectious diseases. [. . .] 'We can't stop COVID-19 without protecting health workers first', said WHO Director-General Dr Tedros Adhanom Ghebreyesus.[16]

The problem was the *pan* in pandemic. COVID-19 was affecting almost everyone, everywhere.* Consequently, staggeringly large volumes of personal protective equipment (PPE) were suddenly needed in all parts of the world simultaneously. At that time, most of the world's PPE was being made in one country: China. Understandably, the Chinese government wanted to make sure that its own healthcare system – a fantastically complex one serving

* Some of the world's smaller island nations kept clear of COVID. And for a while the leadership of North Korea and Turkmenistan simply decided that COVID didn't exist within their borders: https://covid19.who.int/table?tableChartType=heat.

over 1.4 billion people – had access to sufficient supplies before allowing others to buy.

Even if enough PPE could be made, getting it to where it was needed was no longer simple. Global shipping was in a state of chaos as ports shut down due to sick or isolating dockworkers. At one point, while the pleas for PPE were resonating throughout the news, thousands of shipping containers of PPE were stacked up at ports waiting for weeks to get to where they were needed. Emergency military flights were dispatched to bring back relatively minuscule quantities of headline-grabbingly expensive supplies. And where there actually was sufficient supply for different countries, very rational stockpiling by some hospitals sometimes caused shortages in others.

This was a real manufacturing crisis, with life-threatening consequences.

In hospitals all around the world, there was simply not enough PPE to go round, and in the midst of incomplete information and frequently changing and incoherent safety guidelines, health-care workers felt nervous. Rushed Friday-night email edicts from government departments implored hospital staff to not use PPE *unless they really needed to.* Statements from ministers along the lines of *There's no need to worry; everything is under control* were inevitably instantly interpreted as *You should worry; things are really not under control.*

What could be done about this? The World Health Organization pointed to a simple solution. They asked PPE manufacturing firms 'to increase manufacturing by 40 per cent'.[17]

Sounds easy: just make lots more. But as we know, that can be really difficult to do.

Back in 2020, the firms who made PPE were – given a host of COVID-19 related factors such as high levels of staff sickness and absence – struggling to maintain normal operations, let alone ramp things up by 40 per cent. There was simply no way these regular medical supply companies could rapidly increase production to the levels required.

Then, over the horizon marched an eclectic band of unlikely heroes to the rescue. Having heard the news about shortages of key medical supplies, thousands of non-medical manufacturing firms – more used to making jet engines, racing cars or vodka – stepped forward to offer their services. Their numbers were swelled by a throng of individual volunteers who had access to workshops within their local communities.

Tell us what you want, they all cried, *and we will make it!*

So, with all these thousands of extra hands at work, surely that 40 per cent increase in output could easily be achieved?

Well, almost.

There turned out to be a huge problem of what economists call 'information asymmetry'. This is a fancy (and Nobel Prize-winning)[18] way of describing a situation where one group of people have access to more information than another group. In this case, the enthusiastic volunteers needed to know two things: what was needed and how to make it. Both bits of information were surprisingly difficult to access mid-pandemic.

Why was that? Partly, it was because everything was changing so fast. Our knowledge of the transmissibility and impact of COVID-19 was constantly being updated as the virus mutated. Hospital managers had to respond to those frequently changing rules about what type of PPE was required by those doing certain types of jobs. There were also terrifyingly wide variations in the predictions of the number of patients likely to be admitted and needing to be plugged in to specialist equipment. Those same managers pored over spreadsheets, peered into cupboards and were presented with a very confusing picture. Some things were in plentiful supply, some were rapidly dwindling, and some had been completely exhausted. Replenishment was uncertain: things had been ordered, some had arrived, some were promised to be arriving soon, some never arrived. And this was for thousands of units of hundreds of different products, from the simplest plastic apron to the most sophisticated intensive care machines.

Even if a hospital could say, *Right, we need ten thousand surgical caps delivered within ten days, and our current suppliers can't do this. Can anyone else make these for us?*, someone in the hospital needed to provide the exact specifications for those caps to the enthusiastic volunteers. More importantly, as these specifications were going to be sent to people who had never made a surgical cap, they also needed information on exactly how to make them. Even the simplest medical product needs to follow an auditable manufacturing process. This might seem like overkill for something designed to trap stray hair and dandruff, but entirely reasonable for something as life-critical as a machine to control a patient's breathing.

The result of this information asymmetry led to immense frustration on all sides. Already exceptionally overworked hospital staff found themselves dealing with a hundred and one questions about product specifications they simply did not have the time to answer. Some volunteers shot ahead and produced things that turned out to be completely unusable. Others patiently waited for the specifications but, by the time they had reconfigured their factory to meet the required standards, the need for that product had gone.

And yet despite versions of this frustrating tale being repeated all around the world, there were also stories of success.[19] To understand how these were achieved, we need to meet some of those unlikely heroes who saved the day.

How a community project became a manufacturing hub

In a small pot-holed car park squeezed between two rather shabby university buildings in Cambridge, a bright orange and green sign shines from the grimy brickwork: 'Makespace →' Trying not to slip on the frosty, patched-up tarmac, I follow the arrow around a corner to two large, dark wooden doors. I press the doorbell button and, after a pause long enough to make me doubt whether the button is connected to anything, I'm let in to see a space where something quite amazing happened.

Makespace was founded in 2012 when a group of enthusiastic engineers wanted to create a community workshop for the city of Cambridge. Through a combination of charm, wit and finagling, the founders were able to occupy the former print shop of Cambridge University Press, a set of industrial buildings nestling between the imposing gothic frontage of the Pitt Building offices and the River Cam's mill pond.* Makespace provides space and equipment such as 3D printers, laser cutters and other workshop facilities for its several hundred members, who range from hobbyists to small companies.

When COVID-19 exploded across the news, one of the members, Ward, contacted another, Abi, who was working at nearby Addenbrooke's Hospital. Was anything needed that could be made at Makespace? After pondering many options, the conversation homed in on face shields. Face shields are commonly used in hospitals to prevent bodily fluids from a patient splattering into the mouths, noses or eyes of those providing care. Given the nature of how COVID is transmitted, face shields were thought to be especially useful in all sorts of different healthcare settings.

And they were also simple to make.

However, thanks to their backgrounds – Ward as the former CEO of a medical device company and Abi as a current clinical engineer – they had some essential additional knowledge. They knew of the critical importance of something called 'regulatory acceptability' – could they make face shields that the hospital

* And this was also the former home of our institute.

could accept for use by their staff? Without getting this confirmed, going further was pointless.

The technical specifications for regulatory acceptability are excruciatingly detailed. They are probably buried deep in a government website and have catchy names such as BS EN14683:2019+AC:2019. Abi and her colleagues delved into the science and engineering behind these requirements to understand and manage the potential risks of getting gung-ho volunteers to make face shields. Once this had been done, the Makespace team knew what was needed to transform their inventing shed into a medical supplies factory.

Ward and colleagues then designed and refined an appropriate process and layout for every step of making the face shields. By the end of the first week, a volunteer had even developed software for managing this whole new 'factory' within the Makespace walls. Most importantly, the volunteers documented all aspects of the product and the process. It was this documentation that provided a template for other maker spaces and non-medical companies to accelerate their own manufacturing quickly and effectively, along with the quality management system that would support the safe adoption of the face shields by the hospital.*

Meanwhile, as the Makespace team were ramping up activities to churn out thousands of face shields for healthcare workers in Cambridgeshire, a hundred miles to the northwest in Derbyshire, a local clothing company was taking a different approach.

The management team at David Nieper Ltd felt they were well positioned to help address one of the PPE supply challenges – shifting production from ladies' clothes to surgical gowns was an obvious step.

* And there was even yet another link to my favourite subject of baking. As supplies of materials began to run out, one of the Makespace team worked out that the visor plastic material was the same as that used for the transparent display lids of cake boxes – and thus a new source of supply was found.

At first, the need from local hospitals seemed simple. Not enough single-use surgical gowns available from China? Make single-use surgical gowns locally. But by having an experienced garment manufacturer talking directly to hospital staff, a neat idea emerged. They realised that if they provided *reusable* gowns, they wouldn't need a continuous stream of new single-use gowns. And that turned out to be a much better solution, not just mid-crisis but for a more sustainable future.

Specialist textile suppliers were contacted to see how this could be done. A suitable material was identified that could be washed at 73°C and reused up to one hundred times, thus fitting the specification perfectly. And, just like their fellow manufacturers one hundred miles to the southeast in Cambridge, the team at David Nieper needed to show the hospitals that these new gowns met the technical specifications necessary for medical use.

Once it was clear that this could be done, the skilled workers at David Nieper reset their cutting machines to deal with the characteristics of the new materials, tuned the threads and tensions on their sewing machines, and began production. Within a few months, they were supplying twenty-five hospitals with reusable, high-quality, locally manufactured gowns.[20]

Those two stories from Makespace and David Nieper show that the ability to make things locally when a crisis strikes is rather useful. What's more, the solutions that are developed mid-crisis can be useful in the longer term. That message is about to be reinforced as we head down to central London for the start of a third story: the extraordinary tale of the UK's 'Ventilator Challenge'.

As COVID-19 spread relentlessly west towards the UK in the first quarter of 2020, the images and evidence from nations already infected were terrifying. It seemed that, for a significant number of patients, the virus led to major respiratory problems that required them to be put into intensive care. Each of those patients typically required access to a mechanical ventilator, a complex and expensive bit of kit that automatically supports their breathing.

In the UK, government auditors were dispatched with clipboards to count how many ventilators were available in February 2020. They found just under eight thousand. That seemed like quite a lot until scientists pointed out that if trends continued, in the best case, thirty thousand would be needed; worst-case scenario, up to ninety thousand.[21]

Someone was immediately sent off with the government's equivalent of the company credit card to buy as many ventilators as possible anywhere in the world. But, again, the *pan* in pandemic quickly stymied that idea. There were almost none for sale.[22]

What next? Could more ventilators be made in the UK? The government turned to the tiny number of UK-based ventilator companies and told them to start making lots more. A very logical request but – as we know – not straightforward to deliver. If it's hard to get a factory to suddenly make higher quantities of relatively simple things like aprons, gowns and face shields, you can imagine how hard it is to try to do the same for complex machines like ventilators with their hundreds of precision components, stringent requirements for the highest quality and deadly consequences of getting it wrong.

So, the government turned to an old strategy. They decided to run a national competition to come up with new ideas, just like the original Longitude Prize of 1714 that found a technique for ships to determine their longitudinal position at sea. A similar concept was the idea at the heart of a national call to arms that became known as the UK Ventilator Challenge.

The logic of the challenge was simple. If you can't buy any, and current manufacturers might not be able to suddenly make lots more, why not ask engineers to come up with a design for a 'rapidly manufacturable' ventilator? Something that could do the job, but which was specifically designed to be made quickly and easily, and in huge numbers.

The response from the manufacturing community was extraordinary. More than fifty leading firms – some with medical device experience, but most without – looked at what was required and thought, *Sure, let's have a go!* Firms that normally built commercial airliners, vacuum cleaners or Formula 1 racing cars deployed their top talent to come up with new designs.

We should pause here to examine the scale of ambition of what was being asked of these firms. Superficially, a mechanical ventilator doesn't look that complicated: a box on wheels, some plastic pipes, a display screen, and something to emit occasional and reassuring beeps. But beneath the surface is an extremely complex system of pipes, valves, turbines, gas feeds, sensors, controllers and electronics that all have to operate automatically in perfect synchronicity to keep a patient alive for days, weeks or possibly months. It cannot fail. Ever. Unsurprisingly, just as we saw earlier with medicines manufacture, the process of getting a new medical device approved for manufacturing is complex, lengthy and very expensive. But in the same way that COVID-19 vaccines were developed at super-rapid pace, a similar mindset was applied very successfully to address the ventilator shortage.

The manufacturers – some working in collaboration, others alone – pursued different strategies regarding design and clinical sophistication. Out of eleven new and adapted designs that emerged from the collective brainpower of hundreds of engineers, four devices made it all the way through to approval for manufacturing. And one of the teams then took that approval and went on to make and deliver new ventilators. The best way of presenting the scale of what that team achieved is to let the numbers speak for themselves.

5 – Pre-pandemic, the daily production of ventilators typically produced in the UK.

7 – The number of new ventilator manufacturing facilities set up in factories normally used to build airliners, racing cars and jet engines.

1.5 – The number of weeks it took to set up new supply chains for acquiring 42 million parts from around the globe.

3,500 – The number of assembly workers recruited and trained to make ventilators, most of whom had no prior experience of making medical devices.

400 – The peak daily production of ventilators mid-pandemic in the UK.

From five to four hundred precision medical devices produced per day, to the highest quality, by people who had largely never even seen a ventilator outside of a TV medical drama. As a result, in three months, fourteen thousand extra ventilators were made available to UK hospitals. By any measure, that was an outstanding achievement.

So, the world did manage to hit – and exceed – that target of a 40 per cent increase in PPE output. The supply of ventilators was substantially increased. Healthcare systems did not collapse.

But what do these stories really reveal about healthcare device manufacturing? We rediscovered that the ability to make things locally can be rather useful.

Being forced to make things locally also triggered innovation. By not having access to supplies of single-use gowns from China, a local manufacturing system for reusable gowns was developed – which has the side-benefit of being much more sustainable. By working with companies they had never worked with before, the manufacturers in the Ventilator Challenge learned a whole new set of skills from one another and how to use new Fourth Digital Revolution-esque digital technologies to collaborate at speed and scale.

And what about Ward and the community at Makespace? Their work has demonstrated the effectiveness of what can be grandly called 'distributed responsive local manufacturing capability' – that is, having skilled people and machines nearby able to make useful stuff when needed – in supporting the work of hospitals and the community healthcare system. There is some form of what are variously labelled 'makerspaces' or 'fabspaces' in pretty much every part of the world now, from Aachen to Zanzibar. Many of them played an important role in local manufacturing responses to a diverse range of COVID-19-related challenges,[23] and many now provide a tested, local manufacturing resource able to deliver to the diverse needs of their local communities.

The manufacturing community's response to the PPE and ventilator shortages during the pandemic also reminded us that given an overwhelmingly clear and compelling goal, and access to the right resources, amazing people will step forward to make the seemingly impossible possible. There are thousands of inventors and entrepreneurs, scientists and engineers looking at the world of healthcare manufacturing and saying, *We must do things better!* What should give us all a great sense of optimism is how many of them are actually making these new ideas into reality.

Thanks to new production technologies such as 3D printing and modular factories, and near-ubiquitous digital connectivity, we can now rapidly develop and test new ideas. And thanks to major increases in the availability of venture capital, the best of these ideas can be scaled at speed to address the ever evolving needs of our healthcare systems. Too many cancers being identified too late for effective treatment? Develop a breathalyser for early-stage cancer detection. Surgery becoming too expensive? Develop robots that help surgeons perform super-quick keyhole surgery. Medical supply chains failing after a natural disaster in a remote location? Develop mini factories that can be deployed right where they are required to help those in need. People reluctant to get vaccinated due to fear of needles? Design needle-less injectors. Not just neat ideas – all real, *manufactured* things, making people's lives better.*

As our students headed home in December and relative calm descended on the institute, I tried to draw some conclusions from my exploration of the manufacturing of medicines and medical devices during the pandemic. Something blindingly obvious was emerging: having the ability to make stuff is really important when a major crisis hits and everything goes into meltdown. And that is also quite useful in helping ensure we have a more sustainable, resilient and responsive healthcare system.

But there was one other part of the healthcare manufacturing world that I still needed to explore: the factory-like place we invariably end up when we need to be mended. During the COVID-19 pandemic, these were the places that overnight found themselves having to operate under previously unimagined pressures. Rereading the notes I made at that time when working with our local hospital, I start to feel my heart thumping in my chest.

Running hospitals

From: ________@nhs.net
Subject: Re: help from the University
Date: 20 March 2020 at 15:43:32 GMT
To: Tim Minshall <________@cam.ac.uk>

Tim – could you join Monday's meeting? Details below.

* More information on all of these can be found at www.timminshall.com.

That shortest of emails was received on the same Friday that we shut our institute and shifted 350 staff and students to working and studying from home in response to pandemic restrictions. I have a photo of my colleague Liz smiling as she finished the final sweep of the building and locked the main entrance. We assumed, with wonderful naivety, that this would be for just a few weeks while the spread of the virus was brought under control.

The Monday meeting referenced in that email was with some of the senior management and clinical teams from the local ten-thousand-person teaching hospital. We could not meet in person due to lockdown restrictions, so my colleagues and I were all dialling in from our hurriedly equipped home 'offices' scattered across the city and outlying villages. And it was literally dialling in. The hospital IT infrastructure was under such pressure that using things like Teams or Zoom was not allowed. So I sat hunched at the desk in my son's bedroom with my iPhone on speaker, leaning in to try to distinguish between the multiple voices being channelled from a meeting room somewhere in the bowels of the hospital a mile or so down the road. I frantically scribbled the names I heard, trying to make sense of the jobs and role descriptions being listed, switching to scroll through the hospital website to find photos of the names to somehow strengthen the connection to the disembodied voices.

And then the hospital team shared something that made me stop and stare wide-eyed down at the phone. They had done some modelling of various scenarios of the potential impact of the virus on the hospital, and even the best case was quite alarming. The worst case was just too appalling to comprehend.

They wanted us to help them prepare for that worst case.

At this point, you might quite reasonably be thinking it was a bit odd for a team of manufacturing engineers to be asked for help by a hospital. Hospitals in fact have several features in common with factories. As we've seen, factories take some inputs and do some kind of processing (people follow a method using machines and materials) and ship outputs out the door. If you take a cartoonish view of what hospitals do, unhealthy patients come in, medical professionals perform some value-adding activities (diagnose and treat with medical devices and medicines) then ship happier patients out the door.

On that basis, it's not unreasonable to think that some of the skills of running a factory could also be useful in a hospital.

We quickly shook ourselves out of our stupefaction and continued to try to capture the potential implications of this worst-case scenario. There were so many things that could go very badly wrong if that storyline played out in full, or even in part. Here's one example: if there was a big surge in the number of COVID-19 patients requiring oxygen (which was exactly what was then being seen in China and Italy), would the hospital's supply be able to cope? You might have noticed that next to most hospital beds there is a little outlet valve for oxygen. Most are used very infrequently. What happens if all of them are needed, in all wards, all at once? Terrifying cases were being reported from other hospitals where whole oxygen systems wobbled, the pressure dropping when multiple outlets were used across all wards. The consequences of that happening do not need explaining.

And that was just one of their problems. There were so many others. How could they triage and process a massive increase in the number of patients – some infected with the virus, some not – suddenly arriving at the front door? How could they deal with a dramatic fall in available staff as they too succumbed to the virus? How could staff be rapidly tested to make sure they didn't inadvertently accelerate transmission?* With patient numbers exploding and available staff diminishing, how could the balancing of patients to beds to clinical staff be managed, at speed? Finally, with ultimate grimness, how could they deal with the potential

* This is long before any reliable and quick tests, such lateral flow or PCR, had become widely available.

mountain of bodies that could start piling up if things went really badly wrong?

'So, perhaps we'll stop there and see if you have any questions,' said one of the hospital voices.

Once we had managed to reconnect our stunned brains to our gaping mouths, we asked for a few points of clarification and then, rather limply, said we would have a think about it and get back to them.

Reflecting back, I think we were all experiencing what psychologists call 'cognitive dissonance' – trying to reconcile two conflicting worldviews. Our newsfeeds had shown us terrible things happening in hospitals elsewhere in the world, but this was our lovely local hospital, where many of our children had been born, where we and our loved ones had been cared for. Could the horror story we had just been told really unfold here, just down the road, in sunny Cambridge? And as we slowly accepted that this was a real possibility, was there going to be anything useful we could do?

Our colleagues Tom and Duncan (whom you met in the previous chapter) then did something very sensible. They got us to view the hospital's challenges through the eyes of a factory manager. Doing so made us realise that the problems we had just had described to us were actually quite familiar. We were looking at multiple examples of *process instability* often seen in factories. As a factory manager, you want to be sure that all the things that your people and machines are doing (your processes) are operating within an acceptable range of output, speed and quality. You want to know the normal maximums and minimums so that you can spot when something is going wrong and do something about it. It's inevitable that, from time to time, a 'disturbance' comes along and throws one of the processes out of the expected range. This disturbance could be a machine malfunctioning, a person forgetting to do something or a supplier sending the wrong component, which can then have a knock-on effect on other processes, and suddenly your normally smoothly operating factory is melting down.

COVID-19 was the most disturbing of disturbances.

To prepare for this, the hospital team needed to know how COVID-19 *could* affect a range of processes across the whole hospital. To understand the potential impact, someone needed to measure what was happening now – to look at the upper and lower limits of performance of, for example, the oxygen supply, the patient triaging process, the testing process and so on – and produce a model of the likely impact of these limits being exceeded. That would allow frantically busy hospital managers to know at what point they would really need to worry. In a crisis, there's no point wasting precious time worrying when things are actually operating within acceptable limits.

Getting that data, however, was going to be a lot of extra work, and the hospital staff were already flat-out. As with the PPE and ventilator shortages, though, another unlikely group of heroes came to the rescue. Due to the newly imposed international travel restrictions, many of our institute's graduate students were unable to return home. They leapt at the chance to volunteer to do something to help the local community in the imminent battle with COVID-19.

By doing exactly what they had been trained to do in factories, the students looked, listened, questioned, measured, analysed, visualised and presented back to the hospital team. By thinking of the hospital as a factory-like operation, the students were able to give the hospital management team a clearer picture of the actual performance of their processes so that they could better focus their efforts on the right things at the right time.

The COVID-19 pandemic increased the visibility of how hospitals sit in relation to the manufacturing world. From one standpoint, hospitals are on the customer side – they use medicines, supplies and devices made by external manufacturing organisations. Yet hospitals are also themselves manufacturing organisations. They are not only factory-like in how they operate, but also factories in a very real sense: they actually make things. These range from the standard plaster casts which set broken limbs to bespoke surgical devices produced by their clinical engineering teams. Medical manufacturing firms often locate part of their activities within or close to a hospital to ensure that the development and trialling of new devices and treatments align with the needs of the medical staff and patients. And with the shift to personalised medicine, this co-location of production and consumption is going to be an increasingly common feature of hospitals. For example, in cancer treatments such as CAR-T, cells are removed from the patient's body, shipped off for processing via a specialised cold chain, then returned for reinjecting into the patient. The so-called 'vein-to-vein' time is critical, and anything that reduces that gap or simplifies the supply chain is invaluable. More localised processing of cells could reduce supply chain risks and speed up the process.[24]

At the same time, enabled by near-ubiquitous digital connectivity technologies, hospitals are moving beyond the constraints imposed by physical buildings and creating so-called 'virtual wards' or 'hospitals at home'. These new approaches aim to provide patients with hospital-level care at home, safely and in familiar surroundings.[25]

Now, whenever I cycle past the sprawl of hospital and research buildings down the road from my house, my perception of what I am seeing has changed. I can see these buildings for what they really are: factories manufacturing healthcare, whose boundaries grow ever blurrier as their activities are increasingly merged into our local communities.

Speaking with those who work in healthcare manufacturing has been incredibly revealing. Their stories conveyed the complex, creative but at times frustratingly rigid nature of our healthcare systems – and how these systems nearly collapsed during a period of unprecedented but oft-forewarned crisis. But these stories have also given us a glimpse into the future for healthcare manufacturing.

And that future is tantalisingly close.

The three themes of this chapter – of medicine development, device manufacture and hospital operations – have a common thread. The shape of the world of healthcare manufacturing is changing, and its boundaries with other parts of our lives are becoming increasingly blurred. This picture is very similar to the ones painted in previous chapters. We are seeing a changing relationship between those who produce and those who consume. Patients, medicines, devices, hospitals have clearly always been interdependent things, but the bonds are strengthening and becoming increasingly two-way. At the heart of this are the roles that you, me and everybody – individually and collectively – play within the healthcare manufacturing industry.

The world of personalised medicine is allowing us to move from 'one-size-hopefully-fits-all' treatments to ones that target individuals, each with their unique physiology, lifestyle and medical history. The blurring of the line between medical devices and consumer electronics devices – like smart watches and fitness trackers – nudges us towards healthier lifestyles. In addition, it allows us to contribute to the oceans of data from which to pull insights, enable early diagnoses and deliver more effective treatments.

The COVID-19 pandemic also revealed the fragilities of relying on a small number of very large factories far away to make essential, basic supplies. Now, there is wider recognition of the importance of having appropriately skilled people and suitable factories locally to provide healthcare supplies. This can ensure resilient sources of medicines and equipment, more local employment and fewer transportation emissions. And hospitals, as customers of manufactured things and as manufacturing organisations themselves, are changing in response to our desire for more personalised, community-engaged and sustainable approaches to healthcare.

Given that our healthcare systems were already creaking with the challenges of spiralling costs and ever increasing demands before the COVID-19 pandemic, this approach is not just a 'nice to have': it *is* the future of healthcare manufacturing.

We are now close to the end of our journey through the manufacturing world, a journey that has sought to answer the questions we set at the start of this book:

> Why is our modern manufacturing world so fragile and so damaging to the planet?
>
> What can be done to make it less fragile and damaging?

We've explored the 'how?' and 'why?' of making, moving and consuming. We've gone on to look at how change happens in the manufacturing world, the ongoing digitalisation of manufacturing and, in this chapter, the increasingly central role of us as individuals and our communities within manufacturing systems. Along the way, we've seen how the barriers separating our 'regular' world from this manufacturing world are gradually dissolving away.

This brings us to the point where we must confront the biggest issue of them all.

How will the manufacturing world ensure that we humans can continue to survive on this planet?

8: Survive

How we are all, as part of the manufacturing world, going to save the planet

In the 1940s, American psychologist Abraham Maslow proposed the concept of the 'hierarchy of needs'.[1] This showed how humans need to fulfil basic physiological needs before worrying about higher needs – there's little point agonising about how to project a more self-confident version of yourself in the workplace when you are freezing on a mountainside with no food or shelter. You need to deal with the basics first. And for most of us, those basics, those *survival needs* are met by things – like a roof over your head, food and clothing – that have been manufactured by someone else.

Unfortunately, the way we are making and moving all those things needed for us to survive is itself – rather ironically – putting our very survival in jeopardy:[2]

> Manufacturing has become the second-largest source of greenhouse gas emissions after electricity and heat production.
>
> Manufacturing produces a phenomenal amount of waste. A significant proportion of the materials we process and the products we make never even get used or are thrown away when they are still perfectly functional.
>
> Manufacturing is one of the biggest sources of pollutants to our natural environment.

To understand how this has happened – and why we continue to let this happen – we need to look at the making of some specific products. Let's start with how three things that address needs sitting at the base of Maslow's hierarchy are manufactured: shelter, food and clothing.

Manufacturing our shelters

I am very fortunate. I live in a reasonably new house, one that is warm and weatherproof. It is made of a combination of materials including bricks, metals, plastics, wood and cement. And that last one is a bit of a problem.

Cement is a wonder-material that, when mixed with gravel to make concrete, allows a huge range of structures from houses to skyscrapers to bridges to be constructed. Thanks to cement we can construct the eight-hundred-metre-tall Burj Khalifa building or the 165-kilometre Danyang–Kunshan Grand Bridge.* Our modern, increasingly urbanised world has largely been built with cement, and it has delivered significant improvements in quality of life for some of the world's poorest communities. In his book *Material World*, Ed Conway talks of how using cement to cover dirt floors reduces the transmission of parasitic infections, and by replacing mud tracks with concrete-paved roads, two things improve measurably for those who live near them: the wages of those who work and the proportion of children who attend school.[3] Cement makes products that make lives better. Concrete has become the second-most consumed material in the world after water.[4]

But this material is also one of the most carbon-intensive and environmentally damaging of all building materials. It generates massive amounts of CO_2 during its production. There is the CO_2 directly emitted during 'calcination', the key process at the heart of cement production; there is the CO_2 emitted indirectly by the generation of 1,400°C of heat during the processing. This all amounts

* It's actually a viaduct but that, technically, still makes it a bridge. And at the time of writing it holds the Guinness World Record for the longest bridge.

to four billion tonnes of cement produced each year contributing around 8 per cent of total global CO_2 emissions.[5] To put that in perspective, that is four times the level of emissions generated by all of aviation.

While I get to live in a strong, safe house thanks to cement, if my great-grandchildren want to live in a similar location in the UK, they are probably going to need to evolve gills – this part of East Anglia is likely to be underwater thanks to rising sea levels caused by climate change, part of which is being accelerated by our prolific – and increasing[6] – use of cement.

Manufacturing our food

In 2011, the UN published a report showing that nearly one-third of all food produced in the world is wasted.[7] When new data was published in 2021, it showed that things had become even worse: the level of global food waste had gone up to 40 per cent of the total produced.[8]

Every year, we produce around 6 billion tonnes of food and then don't use 2.5 billion tonnes of it. How crazy is that? The CEO of any manufacturing business that wasted 40 per cent of its output would be facing some rather awkward questions from their shareholders. And it's not just the waste of materials, energy and effort. Just over one quarter of all global CO_2 emissions are related to the production of food[9] – that means that around 10 per cent of all emissions are generated producing food that is never eaten.

The problem is spread across the whole journey from 'field to fork'. Losses happen at every stage of this journey, from growing and harvesting (28 per cent), to transportation and processing (26 per cent).[10] That leaves nearly half of the waste occurring once we as consumers get involved. Here's a particularly embarrassing bit of data: it is estimated that over a third of fruit and vegetable produce across Europe is discarded for aesthetic reasons. We won't eat potatoes and carrots unless they look 'right'. And then, in relatively wealthy countries, we do something even more bizarre. We pay good money for food, keep it for a while, then throw it away.[11]

In different parts of the world, losses happen at different stages of that journey. In Europe and North America, over a third of the waste is generated once the food is in the hands of consumers. In sub-Saharan Africa and Southeast Asia, there is very little waste at the consumer end; almost all of it is generated during production, distribution and storage.[12]

Here's one final food-related statistic that I think is worth remembering: our global food manufacturing system currently produces enough to feed every one of the eight billion people in the world. In fact, we produce enough to feed ten billion people. Yet one in nine people is undernourished because, in part, our manufacturing system wastes so much of the food it produces and it never reaches the mouths of those who need it the most.[13]

I think it is fair to say that the manufacturing system we have for delivering even basic nourishment for all is not really fit for purpose.

Manufacturing our clothes

The industry that caters for our need to be appropriately clothed is huge. Worldwide, sixty million people are involved in clothing production. If the broader supply chain (including designers, distributors and retailers) is taken into account, it is more like 300 million.[14] And its economic value is also massive, with annual worldwide sales calculated to be $2.4 trillion.[15] If the global clothing industry were a country, it would sit between Russia and France in terms of GDP.[16]

The production side of the industry's activities can be split into two main parts: the making of the materials and the 'cut and sew' activities to convert those materials into clothes. After this, there are all the distribution and retail activities. Given that the absolute scale and boundaries of the industry are hard to nail down precisely, it is also quite hard to work out its environmental impact. But there is no disagreement on one thing: in terms of emissions and waste it is substantial. Here are some figures to support that point.

> The annual greenhouse gas emissions from textiles production (1.2 billion tonnes) exceeds those of all international flights and maritime shipping.[17]
>
> Over one-third of the raw materials in the textile supply chain end up as waste before a garment or product reaches the consumer.[18]
>
> The textile industry is responsible for 20 per cent of all water pollution.[19]
>
> Nearly three-quarters of material produced for clothing – at a rate of one truck per second – ends up in landfill or is incinerated. Overproduction is estimated to be 10–40 per cent every year.[20] Just 1 per cent is recycled for making new clothes.[21]

Beyond this, after the textiles have been produced, the 'cut and sew' activities have been completed and the clothes and shoes are made available to us as consumers, we behave in a rather odd way that adds to the problem.

We seem to want ever more clothes but to wear them less and chuck them away sooner.[22] This is due in part to the emergence of 'fast fashion', ever quicker turnaround bringing new styles faster and with lower prices. From a commercial point of view, this has been great: high sales, good profits and lots of employment. However, as clothing sales have skyrocketed, we have seen a one-third drop in 'clothing utilisation', the average number of times a garment is worn before disposal.[23] It's not that the clothes are wearing out quicker. We just don't use them for as long as we used to. All of those materials, all of that energy, all of that effort to make things that we chuck away long before the end of their useful life.

Clothing is clearly another part of our manufacturing world that really needs a substantial rethink.

You can see the common message emerging from these three case studies. We have managed to create this alarming dilemma for ourselves. Over the last two and a half centuries we have been developing manufacturing systems to support our survival in the short to medium term – by cheaply and rapidly giving us basic things such as shelter, food and clothing. But in doing so we have also inadvertently created a system that is reducing our chances of survival in the long term.

How did we let that happen?

Our very human approach to timescales

It is easy to make all sorts of claims that we care about the longer-term consequences of our actions but – and I am sadly in this category – many of us find the needs and wants of the present and near future so much more compelling. William MacAskill rather neatly explains why in his book *What We Owe the Future*. The problem is that our descendants, for obvious reasons, cannot vote or lobby or engage in politics. They can't write angry letters to newspapers, or whip up a storm on social media, or glue themselves to motorways. Our descendants are, as MacAskill puts it, 'entirely disenfranchised'.[24]

To help enfranchise our future selves and prevent the self-inflicted destruction of our planet, in the 1980s the United Nations decided that it ought to do something. The UN asked the then Prime Minister of Norway, Gro Harlem Brundtland, to form a commission of wise people to ponder what could be done. Four years later she and her commissioners came back with three hundred pages of ideas. Within these, they defined a blindingly simple concept to which the world would need to adhere to avoid planetary destruction. It was the concept of *sustainable development*, defined as 'development that meets the needs of the present without compromising the ability of future generations to meet their own needs'.[25]

Based on that definition, it's pretty clear that our current system for manufacturing all the things we need – let alone all the things we want – is not really doing its job.

The three examples of shelter, food and clothing have shown the huge inefficiencies and unsustainable practices in each of the make, move and consume stages of the manufacturing process.

But if we look at those three examples from a different angle, we can also see that some really neat ideas are being implemented to start addressing these gargantuan problems.

Making things less harmfully and making less harmful things

I want to stay at the base of Maslow's hierarchy to look at some examples of how things are being improved. And, as I am writing this section in late December while selflessly trying to finish the last of the Christmas cake, I'll start with a food-related example.

Between September and March, much of my childhood in rural southwest Norfolk had a sickly-sweet smell to it, depending on the direction of the wind. The source was a big factory five miles away that extracted sugar from locally grown beet. That factory – run by British Sugar – continues to operate today on an even grander scale than it did in the 1970s.

In a six-month period every year – known as the 'campaign' – millions of tonnes of sugar beet are dug out of the ground and transported from over one thousand farms to this factory via a constantly revolving convoy of tractors and trucks. Once at the factory, the beet is washed, shredded and processed to extract the sugar.

In one campaign, three million tonnes of beet enter at one end of this factory and 400,000 tonnes of sugar comes out the other; just under one kilo ended up in our Christmas cake (and some has now been passed via my sticky fingers to this keyboard as I type these words).

What I didn't know until recently was that this factory has also become one of the most sustainable in the world. It has taken an extreme 'zero-waste' approach to its whole operation that would make Taiichi Ohno and Eiji Toyoda – they of 'lean' manufacturing fame – very happy men. And by doing so, this factory has also rather surprisingly ended up as one of the world's biggest producers of medicinal cannabis.

Perhaps I should explain that in a bit more detail.

Writer Jonathan Meades put it well when he noted that in the Fenlands of East Anglia,* 'The horizon is well named. It is constantly, invariably horizontal.'[26] Almost nothing interrupts the view, and as a result, the vast British Sugar plant at Wissington is pretty hard to miss. Even before you see the seven massive sugar silos, plumes of steam act as a giant navigational aid from many miles away. Turning east off the main north–south highway, I wiggle my way through the village of Southery and exit onto the ruler-straight roads that skirt the blackest of peat fields. Reflecting the international nature of those who provide the logistics for

* The Fenlands of East Anglia are essentially a vast, open-air factory. These great tracts of satisfyingly neat prime agricultural land were created by Dutch entrepreneurs and engineers in the 1600s. They applied their domestically honed skills to drain – and keep dry – these English wetlands using a combination of strategically built channels, dykes and wind-powered pumps. The astounding productivity of these artificially engineered farmlands inspired the wholescale industrialisation of British farming.

this industry in the UK, the 'Beet Traffic Only' sign is also given in, amongst other languages, Croatian, Latvian, Lithuanian and Bulgarian.

As I pull up in the car park, my host Gary is easy to spot in his orange hi-vis jacket. Gary is the man who configured and led the transformation of the plant, and he is keen to show me the extraordinary ambition that he and his former colleagues set themselves. And what is so impressive is that they completed almost everything they set out to do.

Gary and his team applied a single-minded approach to the whole operations of the factory to eliminate *all* possible waste. And where waste could not be eliminated, they tried to find some way in which it could be converted into something of value to someone else. Here are a few examples of what they did.

Beet loaded into the trailers of the hundreds of trucks arriving each day straight from the fields will have mud and stones stuck to them. In the past, these were washed off and chucked away. Now, the mud is resold as topsoil for landscaping, and the stones for aggregate. That's 150,000 tonnes of soil and 5,000 tonnes of stones every year now creating commercial value that would otherwise have just been dumped.

By investing in a fermentation and distillation plant, previously unusable by-products from the sugar extraction process can now be used to generate bioethanol. Now, 55,000 tonnes of this are produced annually and shipped off to provide the required 5 per cent and 10 per cent that is added to petrol to make the E5 and E10 biofuel on sale from petrol forecourts.

As sugar extraction requires a lot of energy, the Wissington plant has its own combined heat and power facility. Any surplus power generated (which can be enough to meet the needs of around 120,000 people) is piped into the national electricity grid. And the surplus heat and CO_2 that would previously be pumped into the atmosphere? Well, thanks to the construction of eighteen hectares of glasshouses, that was used to speed up the ripening of tomatoes and provide a nice little side-earner for the business. When changes in the global tomato market meant that this business was no longer economically viable, they switched to growing medicinal cannabis.

The work of Gary and co. provides one illustration of how the 'make' part of manufacturing can – with some lateral thinking – be adapted to not only reduce harm to the environment, but also create new business opportunities. And there are plenty of other people like Gary who are equally passionate about finding ways to manufacture for our immediate basic needs without putting our future in jeopardy.

We saw earlier the huge impact of cement production on global emissions of CO_2. To reduce this, some scientists and engineers are focusing on the choice of input materials. Over half of the emissions come from producing clinker, one of the main ingredients in cement. Blending clinker with alternative materials is being explored to reduce the need for clinker itself.[27] Others are developing some really clever ideas for recycling cement.[28]

Some are taking a different approach, accepting that the cement industry is not going to be able to make fundamental technological changes in the short term. Instead, they are focusing on ways to make small improvements – incremental innovations – to the efficiency of the current cement making processes.

Sounds a bit dull? Well, think about this. If a typical cement factory reduces its energy consumption by just 1–2 per cent, it will reduce its CO_2 emissions by around fifteen thousand tonnes and save around $500,000 per year. By making cement production just a little bit more efficient, you not only reduce damage to the planet, but also save money. With over three thousand cement factories around the world,[29] that's an annual saving of over $1.5 billion and 45 million tonnes of CO_2. And this realisation drove our former student Dan to co-found a start-up to help firms do just that.

Having worked as an industrial consultant after graduating in aero-engineering, Dan decided to return to university to do a PhD. His topic of choice? How to make the cement industry more efficient. Having done this, he realised that a thesis sitting on a library shelf wasn't going to lead to much actual change by itself. So, he teamed up with his co-founders to launch CarbonRe, a firm that utilises data already being captured by individual cement factories via the many sensors that are stuck all around a typical plant. Dan and the team use AI to find patterns in the high volumes of typically quite messy data to spot areas of inefficiency, developing models of more efficient approaches – a small tweak to this flow here, nudge to that temperature there. This way, they reckon they could reduce fuel costs by 10 per cent and fuel-derived carbon emissions by up to 20 per cent.[30] That's an example of how a small company could help make a tangible reduction in the emissions of one of the most carbon-intensive industries in the world.

And what about that other incredibly polluting part of the manufacturing world that gives us such an immense variety of low-cost clothing but also contributes over a billion tonnes of CO_2, generates mountains of waste material and degrades the natural environment? Reassuringly, efforts are now being directed towards clothing us in a less planet-destroying manner.[31]

To illustrate what can be done, I want to focus on something we produce over three billion pairs of every year and which, at any given moment, some anthropologists estimate that half the world's population are wearing: jeans.[32]

Worldwide sales of jeans generate well over $60 billion per year.[33] The UN estimates around 7,500 litres of water are used in the complete process from field to sale of making a single pair. That's the amount of water that the average person drinks over a period of seven years.[34] Because of their iconic status as well as the scale of the market, finding ways to reduce the environmental harm arising from the production and shipping of jeans has become a bit of a focus for the fashion industry,[35] and there seems to be no shortage of creativity being deployed to this challenge.* One notable example is that of the Spanish organisation Jeanologia, set up by Jose Vidal and his nephew Enrique Silla. They decided to focus on the highly water-intensive and pollution-generating 'finishing' end of the process to deliver the ambitious goal of the 'total dehydration and detoxification in the world of denim'.[36] They developed systems for using ozone instead of chemical bleaching to fade the jeans, meaning fewer harmful pollutants being pumped out, and a 'nano-bubble' system to soften up the jeans and prevent shrinkage, reducing water consumption compared to other methods by over 90 per cent. And for those who want jeans with that distressed look, industrial sandblasting and laborious hand-sanding is replaced by some rather cool laser technology that can create a bespoke vintage look for each and every pair.

Jeanologia have been very successful. It is estimated that one-third of the five billion pairs of jeans produced each year are now made using processes that incorporate their technologies. And again, that's just one company – the good news is that there are many more like them, innovating across supply networks to make the whole textile and fashion system far more planet-friendly.

These stories of the production of sugar, cement and jeans give a glimpse of some of the ways in which the 'make' part of manufacturing is being made less harmful to the planet.

But what about addressing the problems caused by the 'moving' of things into and between the factories, and from factories to

* If you want to read more about the whole complex process of seed-to-shop for denim, I'd recommend starting with *Fashionopolis* by Dana Thomas (2019, Apollo).

customers? Here, there are also positive steps being made to reduce the distances travelled and, where moving is deemed necessary, to find cleaner ways to move things. Let's stick with the same three examples of construction, clothing and food to see how 'moving' is changing.

Move less, move cleaner

If you look at any major construction site, even a very well-managed one, there does seem to be rather a lot of stuff lying around, exposed to the elements – blocks and panels, windows and doors, scaffolding poles and clamps, pipework, lengths of metal, planks of wood, giant bags or piles of sand and gravel, and much, much more. Compared to what you see in, say, a well-managed automotive, electronics or food factory, a construction site seems a bit more 'organic' in its organisation. Things are scattered about, and lots of supplies are coming in from lots of different suppliers, to be used by workers typically working for lots of different sub-contractor organisations. One result of this is waste. A lot of waste. It has been estimated that 13 per cent of material delivered to construction sites never gets used.

So, why not simplify things and get the key parts of buildings assembled in well-organised factories – where things can be more tightly controlled and kept out of the rain and mud – then shipped to the building site for assembly as and when needed? This is known as 'off-site manufacturing'.[37] This approach has been shown to reduce the emissions from transport, with fewer trucks involved given that only large, pre-fabricated units are moved, rather than all the individual bundles of components and materials on multiple trucks. It also has the added benefit of leading to reductions in both time and material waste.[38]

Construction is taking the 'move less' approach by moving fewer, bigger things.

For food, there are some extraordinary things happening that have the potential to dramatically reduce moving distances. Given that most of the world's population lives in cities,[39] one of the most remarkable examples can be seen in the development of technologies to support urban farming.

Urban farmers have come up with some very high-tech solutions to deal with the lack of land, one of which is 'vertical farming'. Here, crops are grown in vertically stacked layers in indoor environments – warehouses, disused offices and factories – where nutrients, humidity, temperature and light can be precisely controlled. The result is localised food production with fewer emissions from transport, less in-transit waste, less use of water, very high yields, and year-round availability of produce. And because of that controlled growing environment, pests and hence pesticides can be avoided. The whole process can be highly automated, reducing the reliance on low-paid, low-skilled labour. For population-dense locations like Singapore and Hong Kong, this makes particular sense, and companies like VertiVegies and Farm66 have squeezed themselves into any space they can find to exploit this opportunity. At the time of writing, vertical farming is a $5 billion industry, projected to grow to $30 billion over the next decade.[40]

A similar logic – make the thing closer to where it is being consumed – is being applied to clothing. As we saw in the previous chapter, Makespace and David Nieper stepped in to make face shields and hospital gowns when COVID-19 disrupted global supply chains.* Moving production closer to home to increase the reliability of supply has the additional benefit of

* The benefits of local approaches to clothing manufacturing can also be seen in initiatives such as Community Clothing, whose story is described in Patrick Grant's book *Less* (2024, William Collins).

moving goods less far, hence reducing the environmental impact of these supply chains.

Sometimes, of course, things do have to be moved over long distances. We've seen how clever things are being done to lessen the environmental impact of land, sea and air transportation. Some of these developments are having a rapid (if long-overdue) impact – such as the increase in sales of electric vehicles. But in 2024, manufacturing firms largely only use electric vehicles for 'last-mile', local delivery of products, and for moving materials and works-in-progress around within the factory. We still seem to be a long way from seeing zero-emission trucks, planes and ships operating at any meaningful scale.

Progress is being made in the freight industry, but it is painfully slow.[41] Just like the cement industry, it is really hard for such a big, complex, highly optimised system to change.

We can now see that things are being done to make the making and moving part of the manufacturing world more sustainable. What about the third bit of manufacturing, the consumption part where we are all involved?

Can we consume things in ways that are more sustainable – or at least less harmful?

From a linear to a circular economy

We all now have access to staggering amounts of data.* The zettabytes of data captured from all stages of make-move-use can now

* Just to remind you of the figures from earlier in the book: Approximately 328.77 million terabytes (or 0.33 zettabytes) of data are created *each day*. Furthermore, over twenty-four hours, 333.22 billion emails are sent, and, at the time of writing, it was estimated that 90 per cent of the world's data was generated in the previous two years.

be employed by manufacturers and consumers to make decisions that have more planet-friendly outcomes. Thanks to all this data, and smart people explaining what it all means,[42] we are all now much better informed about the impact of different manufactured products on the environment.

And it has made us realise something very simple but very profound: we need to be more circular than linear.

Ellen MacArthur is a remarkable individual. Having broken the world record for the fastest solo circumnavigation of the globe by sail in 2005, she described the extraordinary effect this achievement had on her understanding of our impact on the planet: 'No experience in my life could have given me a better understanding of the word "finite".'[43]

Spending seventy-one days alone at sea, relying entirely on the limited food, fuel and other supplies she packed on board, led Ellen to the realisation that 'our global economy is no different – it relies completely on the finite resources we extract, use and then dispose of'.[44] With this mindset, she decided to do something practical, something that could be done at scale to make a fundamental change in how things are made and consumed. She retired from competitive sailing to establish a foundation that helps organisations reduce their negative impact on the planet. The foundation is based on a simple premise:

> In our current economy, we take materials from the Earth, make products from them, and eventually throw them away as waste – the process is linear. In a circular economy, by contrast, we stop waste being produced in the first place.[45]

This simple yet powerful concept of the circular economy builds upon the work of chemist Michael Braungart and architect William McDonough, who first published their ideas in the 2002 bestseller *Cradle to Cradle: Remaking the Way We Make Things*. Their key message was that circularity must keep humans at the centre, and that it must move us from just struggling to deal with an impending crisis to a situation where we are naturally working within the constraints imposed by our planet's finite resources.

So what does this circular approach to making, moving and consuming things look like in practice for manufacturing firms? At its heart is a set of activities with the prefix 're-', with the primary ones being *reduce*, *reuse* and *recycle*. And the order of those words is very important.

Ideally, you want to cut down the amount of things you are making and using (reduce). Then, once its primary purpose has been served, you try to use it for something else (reuse). Finally, as a last resort, you send it to be recycled and its materials used to make new products (recycle).* In a true circular economy, there is no 'waste' at the end. What was previously thought of as waste is now an input to another manufacturing process.

And this is what was powerfully illustrated in the example of Gary and his colleagues working at the British Sugar beet factory. Almost everything in that factory that was regarded as waste (mud, stones, post-processed beet solids and liquids, excess heat, CO_2) now has value for some other process or product – be it garden topsoil, bioethanol or medicinal cannabis production. Thanks to exemplar projects like the Wissington factory, more manufacturers are seeing the commercial advantage of moving to a more circular model.

* But to do so in a way that ideally minimises 'downcycling' – the deterioration of performance each time something goes through a recycling process. While some materials – such as glass and paper – can be repeatedly recycled, plastics degrade each time they pass through the recycling process. As a result, PET plastics used for food and drink containers get downcycled into lower-value materials such as fibres for clothing or carpets. Ideally, it'd be great if things could be 'upcycled' – recycled into things of higher value.

Reduce, reuse, recycle. Three things that are catchily easy to remember, difficult to do consistently and effectively, but which have now become common currency for manufacturing firms. The further we progress with the adoption of these concepts, however, the more complicated things become. We no longer have just those three 're-' words. The list has expanded to include refuse, rethink, redesign, renovate, repair, repurpose, refurbish, remanufacture, recover, regenerate – and so on.*

Examining just three examples from that expanded list can reveal some developments in manufacturing that have the potential to reduce – and possibly reverse – the harm we are doing to our planet.

Renovate

Carl Elefante, former president of the American Institute of Architects, made a simple but powerful statement about how we can make construction more sustainable: 'The greenest building is the one that already exists.'[46]

As we saw earlier, concrete is the most-used material for construction in the world, and its production is one of the most damaging to the environment. Thanks to start-ups like CarbonRe and research teams across the world working with the big cement-producing firms, production is *starting* to be made more efficient and thus less emissive. However, significant changes within this vast, complex global industry are not going to happen overnight. Constructing large new buildings is still going to rely on cement in some shape or form, and, no matter how efficiently it is made, it will contain a lot of embedded carbon.[47]

However, the environmental impact of construction is not just about cement.

* I've given some more info on each of these, along with examples, on this book's website, www.timminshall.com.

Globally, the construction industry also consumes 26 per cent of aluminium output, 50 per cent of steel production and 25 per cent of all plastics.[48] The production of each of those materials, of course, consumes energy and generates emissions. But as Will Hurst, editor of the *Architects' Journal* points out,

> One reason construction consumes so much is because it is based on a wasteful economic model which often involves tearing down existing structures and buildings, disposing of the resulting material in a haphazard fashion, and rebuilding from scratch. According to the Department for the Environment, Food and Rural Affairs (Defra), of the 200 million tonnes of waste generated in Britain annually, 63 per cent is construction debris. We lose more than 50,000 buildings through demolition every year and, while more than 90 per cent of the resulting waste material is recovered, much of this is recycled into a less valuable product or material, rather than being reused.[49]

You might argue that perhaps a new building will be more efficient to operate and so, in the long term, better for the planet, but the timescales for realising that benefit seem to be quite long – 'it can take between 10 and 80 years for a new, energy-efficient building to overcome, through more efficient operations, the negative climate change impacts that were created during the construction process'.[50]

There is a growing view that retrofitting old buildings rather than knocking them down and building anew is generally a more planet-friendly approach to construction. 'Retrofit of existing buildings . . . is cost-effective and generally less controversial, because it conserves and enhances existing places and neighbourhoods. As for carbon emissions, retrofit makes sense because of the substantial embodied energy savings made in repurposing existing buildings, compared with the ultra-high embodied energy costs of demolition and rebuild.'[51]

To help support organisations who want to adopt such approaches, examples of success can be powerful, and there is one here in Cambridge that exemplifies what can be done if a lot of very persuasive people align their efforts on a clear goal. The University of Cambridge wanted to use a six-storey former city centre telephone exchange building 'to show that retrofitting an existing building could be delivered at a cost competitive to a conventional office refurbishment and wanted to maximise recycled and reclaimed materials and minimise negative impacts on the environment and energy use.' And that's exactly what they did. You can read all about the whole process – including the 'warts and all' story and detailed data – on the Entopia Building website.[52]

And who is the tenant in this building? It is the entirely appropriate Institute for Sustainability Leadership.

That example of leading-by-doing brings us on to the next 're-' word, one that is a close cousin of renovating and retrofitting, and one that has become a bit of a complex topic for manufacturers.

Repair

Henry Ford made what now seems like a rather surprising statement in 1922. He stated, 'We want the man who buys one of our products never to have to buy another.'[53]

Sadly, one rather fundamental issue seems to have stymied his remarkably utopian and planet-friendly approach to manufacturing: the economics of corporate growth.

Imagine the scene. Henry Ford is dead pleased at having shown himself to be a manufacturing hero of his age. Driven by a burning desire to provide affordable and reliable personalised transport to the masses – that could be maintained and repaired by the individual owners – his assembly lines have been churning out Model Ts at a phenomenal rate. Then, having said those remarkable words above, his colleagues quietly take him to one side and point out that encouraging each customer to only ever buy one car might not be terribly good for business in the longer term. Telling them to just keep repairing the one they had was a bit commercially short-sighted. Would it not, they might have suggested, be better to persuade customers to buy new ones on a reasonably frequent basis? He clearly acquiesced and, as a result, Ford's company went on to become the commercial success we now know.

But how did Ford – and almost all other manufacturers – make customers believe that products they currently owned were no longer good enough and needed regular replacing? For any rational customer to make such a decision, one of two things needs to happen: either the product must no longer work properly, or there must be a new product on the market that offers some new benefit over the old one.

When technology is developing apace, with each year seeing big improvements in performance, persuading people to replace old for new is relatively easy. Think back to the examples from the early days of cars. When a manufacturer's new model offered an electric starter-motor rather than a shoulder-dislocating, thumb-snapping crank-handle manual one, customers didn't need much persuasion to buy the new model. But what happens when the new product isn't functionally that different or better than the old one?

You could find a way to make your products somehow stop working – or at least not work very well – after a period of time, and also make them hard to repair or upgrade. You could make them *functionally* obsolete (they stop working) or *economically* obsolete (too expensive to repair or upgrade).

It was in that second category that my phone found itself around the time I was writing this chapter. I couldn't bring myself to pay

Apple the cost of replacing the battery in my phone. I wanted to see if I could do something about this myself, but a colleague, who I shall refer to as 'John',* told me the tale of the rather expensive consequence of his attempts to replace his phone battery. Mid-repair, he managed to puncture the battery, triggering one of those very scary 'thermal events'. His unsuccessful attempt to speedily dispose of the smoking, then flaming, mess resulted in his living room needing to be recarpeted. After hearing this, to my shame, I decided to head to the Apple Store to let the professionals deal with this.

However, the fact that I had the option to do the repair myself is really important. Despite resistance from some manufacturers,† the 'right to repair' is increasingly being mandated by governments around the world.[54] And to support those who wish to exert their right to repair, we are seeing more organisations like the Open Repair Alliance, who share data on how to repair electronic devices, and the Repair Café foundation, who support those who want to run local repair events. Other start-ups like iFixit and DirectFix offer parts, tools and support to help repair an increasingly wide range of products.

This is all wonderful to see, but I can't help thinking that this just looks like a successful skirmish in a much longer battle to change our collective thinking in regard to how we engage with products over their whole functional lifetime.

* Because that's his name. He told me there's no point trying to anonymise things as anyone who knows him and reads that story will know who I'm talking about.

† If you start tinkering around with your phone, there's probably not too much harm you can do (but see previous note). However, take a spanner to something like one of those terrifying robotic tree harvesting machines we saw at the start of the book, and there is significant risk of harming yourself and those around you. That is one of the reasons why some companies will still insist that all fixes are only done by authorised repair shops where the quality of the work and the parts can be ensured.

Part of the problem is something that goes beyond the rational and into the emotional.

Looking back to the 1920s, US academic Aaron Perzanowski noted that 'conventional wisdom held that "changing design so that old models will become obsolete and new ones have the chance to be bought" was simply 'good manufacturing practice" and "clever business"'.[55] Some even blamed the slow exit from the Great Depression of that decade on consumers selfishly not replacing old with new, fuming that '(p)eople everywhere are today disobeying the law of obsolescence,' and arguing that the government needed to step in to force people to buy new things.[56] As it turned out, no such drastic intervention was needed. Instead, manufacturers found ways to make us want new things even when the things we already had were perfectly good.

Manufacturers deployed *psychological obsolescence*. They realised that, as consumers became more affluent, they could shift their marketing efforts from the lower end of Maslow's hierarchy to the higher layers, where we worry about how our purchases send signals to people we want to impress. One way in which manufacturers did this was by introducing annual models, which – often through easily visible feature changes – showed to our peers that we owned the latest model.

Smart manufacturers then also found ways to link the emotional and the rational by combining functional, economic *and* psychological obsolescence. Not only does your phone look old; it can't use the latest apps, and the battery struggles to last half a day. And no one wants to feel like a twit standing at the train ticket barrier or supermarket checkout, nor do they want to be loomed over by a frustrated waiter while you mutter how your phone should definitely still have more than 1 per cent power left.

This reframing of our relationship with products throughout their whole life leads us neatly on to the final 're-' word.

Rethink

Earlier in this book, we looked at the idea that 'no one wants a drill. What they want is the hole'.[57] This is why airlines often pay engine manufacturers for a certain number of hours of flight per year, rather than buying the engines and maintaining them themselves. These 'power by the hour' and similar 'pay-per-use' approaches used elsewhere not only make commercial sense but can also deliver a positive impact on sustainability. Manufacturers have a strong incentive to ensure that their products are as efficient and durable as possible, because they will bear the maintenance and repair costs. An unintended consequence of this 'service-based' approach for the planet? More reliable and long-lasting products with reduced need for replacement of spare parts that consume energy and resources to make and install.

This rethinking by manufacturers of how they deliver what the customer really wants – and the benefits this delivers for the planet – is now being seen in many other sectors. For example, research shows that each car used by 'car share clubs' like Zipcar and Getaround removes the need for around twenty-three individually owned cars (which, typically, spend 96 per cent of their time not being used).[58] In clothing, renting rather than buying clothes has become a thing. Research commissioned by one of the companies doing this, Rent the Runway, estimates that its business approach 'has displaced the production of 1.3 million new garments since 2010 – leading to savings of 67 million gallons of water, 98.6 million kWh of energy and 44.2 million pounds of CO_2 emissions over the past decade.'[59]

Rethinking our approach to how we make things is fundamental to reducing or reversing the harm we have done to the planet, and yet we can only go so far with tweaking this element of the system. As the examples above have shown, sometimes a completely new approach is the only effective way forward.

Why aren't we doing this faster and at a global scale?

We now have the data showing that the old linear approach to manufacturing of *make-use-dispose* cannot continue. Manufacturers no longer need to be persuaded of the need to do less harm to the planet: it is now embedded at the core of almost all businesses. And, thanks to the likes of Ellen MacArthur and many others, we have the tools for governments and organisations to transform to more sustainable, more circular ways of making, moving and consuming.

In many countries, the impact of using this data and these tools has already led to an impressive drop in harmful emissions from manufacturing,[60] but we are still far from where we need to be. Manufacturing is still one of the top three biggest generators of CO_2 and creates (literally) unsustainable amounts of waste. Things definitely are changing for the better, but there is much more to be done, and it needs to be done much quicker.

So, you might be wondering, if we know the things that need doing, and how to do them, why aren't they being done 'at scale'?

It comes down to three things.

1. Whose problem is it?

A few years ago I saw a graph displaying some good news about the amount of CO_2 being generated by UK manufacturing. The data showed a huge reduction in emissions from manufacturing activities in the UK since the 1980s. This seemed like a very good thing until it was pointed out to me that a big chunk of this reduction was the result of shutting down factories in the UK and moving production activities to other countries, mainly China. Emissions were still being generated by these manufacturing activities, but the UK could proudly boast that it had done an amazing thing by ensuring that its *domestic* manufacturing activities were less harmful to the planet.

This is the nub of why we aren't moving fast enough to reduce the harm caused by manufacturing. It comes down to how manufacturing emissions are split into three different 'scopes'.[61]

It starts simply enough. Scope 1 covers the emissions generated *directly* by what you do. Whether you are making cakes or rockets, it is the same. Scope 2 is about the emissions generated *indirectly* by you. For example, unless you run your own power station, the electricity used in your factory will come from the grid, and its generation will have generated some emissions – the proportion of emissions generated in making electricity for you will fall under Scope 2. A bit trickier to measure, but still doable.

But then we have Scope 3, and things get very messy.[62] Into this category goes everything not covered in the first two categories. It is all the emissions generated across your supply network, and by your customers when using the product. This is quite tricky to deal with. Why? Because the data on what exactly is going on as you move deeper into complex supplier networks is frankly quite flaky.[63]

Even if you do get good data to pinpoint exactly which activities across your supply networks are particularly bad for the planet, these activities are being done by other companies and people. How do you persuade your suppliers to make sure their activities are more sustainable? If you say to them that they must switch to renewable sources of energy, they might – quite reasonably – point out that this is going to cost them a lot of money, meaning the price they charge you is going to go up. You, as a manufacturer, will have to either absorb that cost yourself or pass it on to your customers.

The worrying thing is that most of the emissions from manufacturing – more than 70 per cent – fall into this third category.[64]

Similar categorisation of the problem can be applied when we look beyond emissions to other environmentally damaging aspects of manufacturing. Recycling is a good example.

We could insist that the organisation that has manufactured

a product should also be responsible for the recycling once the customer has finished with it. However, in most cases, this isn't what happens. Think about the recycling of glass bottles. I buy a bottle of beer, enjoy consuming its crisp, refreshing contents, and then want to recycle the bottle. Neither the brewery nor the retailer pays for that to be done. The recycling is often paid for and managed by local councils.

There are examples where this is changing. For example, as a result of changes in legislation, manufacturers and retailers now have more responsibility for the recycling and disposal of electronic products that they have made and sold.[65]

Assuming that we can decide who is responsible for reducing the environmental impact of the making and moving of a product and persuade them that they need to do something about this, that leads us to the second problem.

2. How to manage trade-offs?

As we have seen, three questions represent very common manufacturing trade-offs:

How good does the customer want the product to be? (quality)

How quickly does the customer want it? (time)

How much resource will this need? (cost)

The customer wants it quicker? That'll mean it's going to cost more, or not be as good. The customer wants it at a higher quality? It'll take longer or cost more.

Managing trade-offs becomes very complicated when considering how best to increase the sustainability of manufacturing.

There is no shortage of things we know we could do to reduce the environmental impact of our making, moving and consuming activities. But which ones are 'best'? What is the 'right' thing to do?

If you as a consumer want to buy clothes that have been made in a way that minimises damage to the planet, it might seem quite straightforward, but going back to the example of jeans illustrates that this is very much not the case.

Let's try to bring in some of the wide range of variables that need to be traded off against each other to make sure we buy the 'most sustainable' jeans. We need to include an understanding of where and how the cotton was grown. Is the cotton organic, 'conventional' (non-organic), or grown adhering to the rules set by the 'Better Cotton Initiative'?[66] How was the cotton processed? Was natural indigo or chemical dye used to add the colour? How was waste water processed at the mill? How far did the rolls of prepared denim travel from mill to factory? What were the working conditions in the mill and in the jean factory? What approach was used for finishing? Were the more sustainable Jeanologia techniques used or the more traditional, water-, chemical- and labour-intensive methods? Then there is the issue of how you choose to purchase, maintain and dispose of the jeans. And, of course, you need to think about whether it is better to buy new or second hand, or even just rent.

All that generates a pretty complicated decision-making algorithm – and that's just for a single pair of jeans. Imagine trying to scale that level of analysis across the whole baffling complexity of all manufacturing systems and you get a flavour of the challenge of trying to manage sustainability trade-offs.

If we assume once again that we are still committed to driving up the sustainability of our manufacturing systems, that we can identify where 'ownership' of the problem sits, *and* that we are comfortable with the trade-offs we need to make, that leaves us with one final problem.

3. That old problem of inertia

Throughout this book, we've discussed how the complex system of global manufacturing is affected by shocks – such as a pandemic, or a ship getting stuck in the Suez Canal – and how it can recover its balance. But we've also seen how trying to adapt the system in response to a longer-term change – such as that triggered by the arrival of a new technology – is really challenging. Our transition from internal-combustion-engine-powered vehicles to electric vehicles illustrates the difficulty of changing just one – albeit very significant – part of our manufacturing system.

But transforming *all* parts of our global manufacturing system so that it becomes less harmful to the planet is a project of immense scale. It requires the introduction of a completely new order of making, moving and consuming. And, in the words of Florentine diplomat and philosopher Niccolò Machiavelli, 'there is nothing more difficult to take in hand, more perilous to conduct, or more uncertain in its success, than to take the lead in the introduction of a new order of things'.[67]

Aside from the sheer scale of the task, there is yet another problem. Things need doing in collaboration with others. No single producer, supplier or consumer can make substantial changes to the system. Individual manufacturing firms and people can act as exemplars, vanguards of change, but the transformation of a complex system requires people and organisations to work in partnership. And moving to more circular approaches to manufacturing illustrates this very well.[68]

For Gary at British Sugar, it's all very well him wanting to use excess heat and CO_2 to help grow tomatoes and medicinal cannabis, but someone's got to build and manage the greenhouses and run a whole business of selling these non-sugar-related products.

Circular manufacturing requires the synchronisation of different business activities, often across different markets. For example, as demand for newsprint has fallen, the amount of paper available for recycling has also fallen. So, paper companies must look more widely for sources of recycled paper. More diverse sources of recycled paper might mean more variation in quality, which could require adjustments to the design and operation of the rolling mills. And as demand increases for more paper to be used in other parts of manufacturing, like packaging – responding to our desire to use less plastic – this drives up demand for recycled paper, creating more competition in the market, and more complexity.

What about moving all of us from simple consumers to active partners in the circular manufacturing system? Yes, we can be told that we should recycle more, but as the earlier example of recycling bottles showed, the manufacturers need to work with us to make that happen, not just devolve responsibility to us or to public service providers.

Ownership of the problem, managing trade-offs, and trying to overcome the inertia inherent in changing a complex system: three pretty powerful reasons why our manufacturing world is not moving fast enough to achieving that UN goal of meeting 'the needs of the present without compromising the ability of future generations to meet their own needs'.

So, what do we need to do?

> Anyone who thinks that you can have infinite growth in a finite environment is either a madman or an economist.[69]

Economic growth has become something that we take for granted, but there have long been concerns at how the exponential level of growth experienced since the First Industrial Revolution can be maintained. Thomas Malthus made the gloomy prediction in the late eighteenth century that unconstrained growth leads us to a dangerous place. As nations become more prosperous through the exploitation of resources, they grow. But Malthus prophesied that as resources such as food and fuel are finite, growth will inevitably have to slow down or very bad things will happen – like famine and wars.

As our understanding of the issues has improved, versions of this message have developed over the years, shifting in tone from *We are all doomed!* to *We are all doomed unless we do something!* to *These are the things we need to do to prevent us being doomed!* and more recently and helpfully to *These are the things we need to do and this is how to do them.**

Part of this trend towards more positive thinking has arisen by us having access to much better data, which has driven higher levels of awareness, propelling us to want to develop solutions.

But one amazingly unhelpful three-letter acronym continues to slow our progress.

GDP – gross domestic product – is one of the most commonly used measures of how an economy is performing. It can be measured in different ways,[70] but fundamentally it's based on how much money individuals, businesses and governments are spending. From one perspective, this is logical, simple and helpful. If consumers are spending more money, companies are investing more in developing their businesses, and governments are putting more money into developing our public services, that's wonderful. However, there is so much that GDP does not take into account, such as the health, education and well-being of the population – and the impact of what we do on our environment.[71] As Diane

* But to do this requires companies to be willing to share the details of what they have done. The alarming phenomenon of 'Green Hushing' – where companies opt to keep quiet about the good things they have done in case this attracts attention to areas where they have made less progress – risks putting this ability to share innovative approaches to sustainability in jeopardy.

Coyle from the Bennett Institute for Public Policy puts it, 'A focus on GDP without proper regard for environmental degradation or inequality has been a disaster for global ecosystems and undermined social cohesion.'[72]

I can give you two quick manufacturing-related examples to illustrate the bizarre outcomes that can arise from an over-focus on GDP. In 1989, the oil tanker *Exxon Valdez* crashed into the coast of Alaska creating what is regarded as one of the worst environmental disasters of all time. Viewed in terms of GDP, however, the Alaskan economy did very well as huge numbers of people rushed there to help with the clean-up, spending loads of money on hotels and food.[73] Alternatively we can look at the ongoing deforestation of swathes of the Amazon to allow for the development of grazing land for cattle.[74] It is great for GDP, as spending within the forestry, timber and ranching industries goes up.

If governments remain fixated on measuring national success by increasing GDP,[75] transforming our manufacturing system to one that is more sustainable is going to be an uphill battle.

One idea that could help to move things forward is that of 'de-coupling'.

This refers to 'breaking the link between "environmental bads" and "economic goods"'.[76] In other words, focusing on finding ways for your economy to grow but 'de-coupling' this from the need to do environmentally bad things like burning more fossil fuels. Yes, we want economic growth, but we don't want this to happen while increasing emissions. Having this concept front-and-centre for our politicians is really important if we are to accelerate the changes that are needed to the manufacturing world.

And there is some good news. This de-coupling is happening. There's a simple graph you can look at on the 'Our World in Data' website. For the UK, it shows two lines over the period 1990 to 2022: one going up by 55 per cent (GDP per capita), the other going down by 55 per cent (CO_2 emissions per capita). For the US, the figures are up 60 per cent for GDP, down 28 per cent for CO_2.[77]

Things are definitely heading in the right direction. We just need to keep them moving that way. And moving a lot faster.

Towards manufacturing that treads more lightly on our planet

If we now apply a 'sustainability' filter to a quick sprint through the manufacturing activities covered in the first half of this book, a rather positive set of stories emerge.

Looking at all the steps taken in manufacturing a product as superficially simple as a roll of toilet paper, you can see the issues we are discussing here writ large. Yes, the whole process of felling, pulping, transporting, rolling, distributing and selling is hardly the most planet-friendly. But things are changing for the better. There's been an increased use of recycled paper as the main raw material, as well as a reduction in the impact of various processing steps such as bleaching, along with some firms switching to more sustainable virgin material such as bamboo.

The principles of 'zero waste' provide an inherently commercially focused way of helping managers minimise the environmental impact of their factories. And this efficiency-driven zero-waste approach to running a manufacturing business is what allows the likes of CarbonRe, Jeanologia and British Sugar to gain traction in helping reduce the emissions and pollution from manufacturing – they are going *with* the grain of what a good manufacturing business should already be doing to remain competitive.

There is also good news when we look at logistics. The growing trend towards reducing the length of supply chains not only reduces the risks of disruptions, it also reduces emissions of CO_2 arising from all that fossil-fuelled moving of things thousands of miles by land, sea and air. And if movement over long distances must happen, new technologies and new fuels are arriving to help make the generation of energy to power the ships, planes and trucks less harmful to the environment.

Looking at the consumption side of things, there is more good news for the planet. The technologies now available to ensure tighter coupling of customer requirements with production capabilities are accelerating us along the path to the nirvana of mass customisation – where manufacturers only make exactly what individual consumers want to buy.

All that being said, the second half of this book has revealed why change at the scale required to bring about Machiavelli's 'new order of things' is going to be very hard. However, the ongoing digitalisation of manufacturing is allowing us to apply frankly awesome levels of computing power to make our manufacturing systems more responsive to our needs and less harmful to our planet.* Rather annoyingly, this is also creating new sustainability challenges as we recognise the harmful impact of running all those digital systems.

And looking at healthcare provides a great illustration of the multiple sustainability-related trade-offs that will face every part of the manufacturing world. By moving to more localised and personalised manufacturing, significant benefits can be realised. But there will always be trade-offs. As patient care becomes more distributed within our communities, we reduce the burden on vast energy-inefficient hospitals. However, with more patient-centric approaches, we are likely to increase the need for single-use devices that can safely be used at home. Levels of emissions will fall, but levels of plastic waste will go up. Our path to a more sustainable manufacturing world will be strewn with such trade-offs.

* Spookily on cue, this alert just popped up as I was typing this paragraph: 'Can artificial intelligence pave the way for greener cement and steel?'

I'd like to leave you with three messages from this survival-focused chapter of our journey through the manufacturing world.

Manufacturing needs fixing and redesigning

Although it is remarkable in so many ways, the way we operate our global manufacturing system is making a significant negative contribution to the health of the planet and its population. So, we must modify the way we operate this system and, in many cases, redesign significant parts of it. For the former, this chapter has shown lots of ways in which such operational improvements can be achieved. For the latter, moving from a linear extract-make-use-dispose approach to a circular model provides a great template for this redesign.

But could we go further?

Much of the discussion about sustainability and manufacturing uses the word 'zero' as the ultimate goal. Zero waste makes sense. 'Zero emissions' sounds good. But what if we could go further? What if, instead of just not making things worse, the making of things could actually make the environment better? What if we could move from 'net-zero' to 'net-positive' manufacturing, where the act of manufacturing actually makes a net-beneficial contribution to the environment? Think: factories whose by-products are purified water or clean surplus energy or nutrients for the soil.

Could our manufacturing world, like the natural world, become *regenerative*?

Sound a bit far-fetched? Well, some people are exploring this idea now.[78] And they are drawing upon concepts from the natural environment, where there is a growing (sorry) movement towards regenerative agriculture.[79] With regenerative manufacturing, the factory would stop being something that we recognised as being important but wanted to be 'away' from our lives. Factories would become something that we *want* to have as part of our communities because of the benefits – cleaner air, purified water, free heat – we could gain from being close to them.

Manufacturing *is* the solution

Manufacturing is not just something that needs fixing. It is the source of the products that will ensure the long-term habitability of this planet. If we want zero-emission transport and energy systems, someone is going have to design and build every part of them. If we want to transform our food and healthcare systems so that they are more sustainable, a whole host of new things need to be dreamed up and produced at scale.

These products will fall into two buckets.[80] There are those to *mitigate* the damage we are doing to the planet – all the systems we need for delivering renewable energy, for example. But we also know that, certainly in the short to medium term, the damage has already been done. So, the second bucket contains all the things that need making to allow us to *adapt* to the problems we have created. This includes the development of systems to help cope with rising sea-levels, droughts or forest fires.

Manufacturing is the mechanism for delivering the systems to mitigate against doing any more harm and adapt to the fact that that the world is going to be very different in the future as a result of our past – and sadly ongoing – actions.

Manufacturing must help us thrive, not just survive

We have mainly looked at how we can ensure that manufacturing does not inadvertently, prematurely terminate our ability to inhabit this planet. But 'doing manufacturing right' can lead to other mutually reinforcing benefits.

For example, if we choose factory locations that minimise emissions from transport, we can also gain the economic benefits of local production. As Peter Marsh, former Manufacturing Editor of the *Financial Times* puts it, 'A vigorous and resourceful manufacturing sector will be vital to any country that wants to be regarded as a twenty-first century success.'[81] This echoes the points noted earlier from economist Ha-Joon Chang rubbishing the argument that most economies can become 'post-industrial' and thrive just by delivering services.

There is another advantage in having manufacturing at the heart of your economy. Relying on single sources of production located far away in giant factories can decrease our collective abilities to respond to crises. Reflecting on the COVID-19 pandemic, Jeremy Farrar, the former head of the Wellcome Trust, one of the biggest UK medical charities said, 'the ability to respond to a crisis is determined by what people, infrastructure, and trust you have in place before it happens'.[82] And a key part of that infrastructure he refers to is the ability to make things – like vaccines, PPE and medical devices. Assuming that 'someone, somewhere else' is going to do that manufacturing for you is all well and good, but what happens when that 'somewhere else' is facing its own crises? You can't magic up the infrastructure, the skilled people, the factories and the supply chains overnight. Manufacturing requires long-term nurturing and support – by both companies and governments.

Manufacturing plays another really important role in helping us thrive on this planet. Manufacturing is all about people. People are at the heart of all manufacturing systems – they design, build, and operate them. When done right, manufacturing offers jobs that are rewarding – in two senses. Data shows that salaries for skilled manufacturing workers sit above national averages,[83] and there are also non-financial, psychological rewards of having a job that is about *making* something.[84] We must not, however, be blind to the fact that manufacturing can also drive people into roles they don't want (children mining for rare earth metals for components for our mobile phones; monotonous low-paid, low-skilled, low-security jobs) or force them out of roles they do want (robots replacing skilled, well-paid humans).

There is a sweet spot for manufacturing systems. When a manufacturing system is well designed, taking a human-centric approach, it can deliver employment opportunities that are both financially and psychologically rewarding. The best manufacturing firms know this and focus not only on supporting individuals to continually develop and refine their skills, but also on providing factory environments that are orders of magnitude removed from the dark, dirty, dangerous ones of . . . well, sadly, many corners of the world today.

And there are even more benefits. By having roles for individuals that are attractive, there are advantages for the local community in which these jobs are embedded. Manufacturing skills take time to develop. They require companies that are committed to developing those skills, and for those working in education to support those firms and their workers to develop those skills. When done well, a virtuous circle is created. By creating a community across which manufacturing jobs are supported, the community itself becomes stronger.*

To close off this section, I want to highlight something really important said by the authors of *Cradle to Cradle*, one of the first books that set out the agenda for transforming our manufacturing system towards planetary good. Our approach should be 'like good gardening. It is not about "saving" the planet but about

* I've put some great examples on this book's website (www.timminshall.com) of how such approaches are successfully revitalising some of our most economically deprived communities.

learning how to thrive on it. [. . .] We cannot move forward if we merely "blame and shame" ourselves and others. We need a spirit of cooperation among ourselves.'[85]

And that's where you come in. We all have a role to play in enabling this new manufacturing system.

In your pocket or in your bag or on your table, you have a mini supercomputer that gives you access to the greatest data resource humanity has ever known. You and I now know so much more about the impact of how we make and move things than those living in pre-internet times, allowing us the chance to make better choices. But as we know from wise philosophers of our age, 'With great power comes great responsibility.'* It's up to us individually to decide how comfortable we are trading off the clear benefits of convenience, speed and low cost against the consequential impacts on the planet.

We started this chapter with a worrying dilemma. The system we have built to provide for our basic survival needs is now destroying our planet. The data and examples shown in this chapter provide ample evidence of the scale and seriousness of this challenge. But this chapter is also about optimism. Manufactured products and systems such as the internet, the billions of IIoT devices and the server farms that power increasingly ubiquitous AI tools now let us see what is happening across the ever more complex networks of production that span the globe. From field to fork, from tree to toilet, from mine to phone, we now have much greater visibility and understanding of how and where the physical things that enable every aspect of our lives are made and brought to us. We have had returned to us the sight we had before the First

* In this case, Stan Lee, author of *Spiderman*.

Industrial Revolution, a time when we had much greater, more intimate connection with the processes of manufacturing. And with that restored vision, we can actively participate in the transition of the manufacturing world through being 'less bad' and to making things much better.

To continue to do that in a way that will really make a difference, we are going to need the most skilled people, the wisest investors, the most visionary politicians and the most scale-up-focused entrepreneurs targeting all their efforts on transforming our manufacturing systems. As I hope you can see from what we've covered in this book, this is now happening – we just need it to happen a lot faster, and at much greater scale.

We are far from out of the woods, but I think we are on the right path. Let's end by heading to another room full of overexcited school pupils to show why I think we have grounds for such optimism.

Epilogue

'Ready?' mouths our host, through the din. My colleague Niamh and I look at each other, then nod our response. Our host walks to the front of the stage, pauses and does something that has a remarkable effect. She raises one hand. The lecture theatre of over five hundred jabbering pupils falls almost instantly silent. Such power. Such a contrast to the chaos of the classroom at the start of this book.

Forty-five minutes later, we are forced to draw things to a close as we are reminded that the buses are doubled-parked outside. We've taken our audience through a very cut-down version of what we've covered in this book. They get it, and they want to know more. Here is a sample of the sort of questions they – and other children at similar talks – asked: When will we get planes that don't pollute the skies? Will this new robot technology make life worse for some people? Aren't factories always bad for the environment? Could we 3D print a car? Can we build a factory in space? Why don't we make electric lorries? Can you help me fix my trainers?

Those seem to reflect a pretty healthy level of interest in the manufacturing world.

As I looked back at my notes from that event, a thought occurred to me. I have no idea when you are reading or listening to this. It might be shortly after first publication in early 2025, or many months or even years later. It could be that, by the time you see or hear these words, some of the challenges that faced the manufacturing world at the time of writing have been solved. It could be that we are now well on the way to ensuring we make the things we want in ways that tread more lightly on the planet, are resilient to crises and support thriving local communities. And it is just possible that some of those children sitting in that lecture theatre described above, or in that classroom at the start of this book, are the very people who helped make that change happen.

Wouldn't that be great? Certainly sounds better than imagining we've learned nothing from recent and ongoing crises. Or that we've failed to harness the huge opportunity of AI as a tool to support human endeavour. Or that all those increasingly self-aware industrial and home delivery robots have realised that they are in dead-end jobs and decided to make us work for them.

Sitting on my kitchen table is a tatty box-file of interesting bits and bobs I have picked up along the way during this journey through the world of manufacturing. I've long given up trying to get the lid to close. There is, of course, the Mr Incredible doll, with his head now reattached the correct way round. There's a prehistoric hard disk drive from the computer that those feral children ripped out of the defunct computer I brought into their classroom. As blood oozes from the small cut I just got drawing my finger over the electrical connectors, I marvel at how no one got hurt in that classroom. I smile as I spot the sticker on the metal casing: 'Made in Malaysia'. The physical map of the global supply network created with those children using the innards of that old computer reappears in my head.

There's a single pristine sheet of toilet paper taken from a posh French hotel. I kept this one as it is so extraordinarily soft, intricately embossed, and multi-layered. You and I now have a pretty good idea of how that product was created, of the extraordinary scale of endeavour to get it from tree to bathroom – the decades of forestry management, the brutal, sudden harvesting, the lakes of water and megawatts of energy used to process the pulp, the thousands of miles travelled by its constituent parts, the scale of the rolling mill converting liquid goop to rolls of perfectly formed soft, strong paper, and the complexity of the retail logistics needed to get that sheet to its point of use.

That sheet now bookmarks a page in my copy of the history of Fitzbillies bakery. Moody black-and-white photos spark the memory of that cold, dark, far-too-early morning when baker Sam showed me how he and his team prepare, bake and ship the daily batches of eight hundred Chelsea buns. Rummaging on through the box, more memories emerge. A pen emblazoned with the corporate colours of Bouygues Construction, a photo of my moped from 1984 (with repaired suspension), a miniature racing car, a ludicrously over-engineered drinks bottle with the Chinese EV firm Zeekr's branding, and a DIY shop flyer printed on paper rolled by Palm Paper. Each is a link to the idea that what you make and how many you make will define the type and layout of a factory, be it project, job shop, batch production, mass production or continuous production.

One of many old iPhones – the older among which now seem comically small – serves as a reminder of the hundreds of thousands of kilometres of toing and froing of materials, components and subsystems across the globe required to make such complex products.

A Post-it note with the scrawl 'Keep!' is stuck on to a copy of *The Economist* with the cover story 'Print me a Stradivarius', weighed down by a small gas-turbine blade I 'borrowed' from a colleague's office shelf many years ago. Those objects show how manufacturers try to match what they make with what we as often irrational and fickle consumers really want. The Post-it is an oft-cited example of customers not knowing they want something until it was put in their hands; the gas-turbine blade is a link to the aerospace industry, where manufacturers try to 'chase demand' – matching the output of their factory to every twitch in the wiggly line of predicted demand. And the 'Print me a Stradivarius' story? That's a reminder of where manufacturing might be heading – a world where you don't need to worry about persuading customers to buy the specific things your factory can make. 3D printing is emblematic of adaptable technologies which could usher in the era of mass customisation.

There are a few ribbons of unseparated bimetallic blades picked up in the Strix factory on the Isle of Man resting on a tatty booklet from the Shell company printed in 1931 that provided valuable insight on the formative years of the petrol-powered car industry. Both link to the challenges of designing, making and growing the market for new products – be they kettle controls or cars. And if I flick my eyes between the car-battery-sized Motorola 4500x phone from the 1980s now wedging open my door and my current smartphone sitting next to this laptop, that's a pretty impressive illustration of the results of another kind of manufacturing change – the relentless incremental improvement of products and processes.

I pick out one more item of debris recovered from the chaotic classroom incident at the start of this book. It's a dusty black object, about fifteen centimetres long, five centimetres wide, five centimetres tall. Most of the upper surface is an array of metal prongs, and at the top is a small electric fan. Along one side are the words 'AMD Athlon™ Processor'. This is a microprocessor from the late 1990s, and the bulk of this object consists of the paraphernalia needed to prevent 37 million transistors overheating while running Windows 98.* The ability to manufacture billions of devices like this one transformed almost every aspect of the manufacturing world. And the speed and scale of development

* And just to show the scale of progress, I remind you that the 2024 Apple M3 Max microprocessor had 92 *billion* transistors.

has been phenomenal. I briefly panic when I can't find a precious object lent to me by another colleague. Then I spot it hidden under a boarding pass: a single IBM punched card from the 1960s, a very tangible and direct connection to the pre-electronic world of computing and digital control kicked off by Joseph Jacquard and his looms back in the 1700s. And that Air China boarding pass transports me back to the visit to the Zeekr EV factory where digitalisation is making a whole manufacturing system – making, moving and consuming – 'smart'.

I sift through the remaining items in the box to find three things linked to the evolving world of manufacturing for healthcare. There's that product that went from specialist and professional to everyday and commonplace: the pale-blue surgical face mask. Its use during the COVID-19 pandemic illustrates so well what happens when a manufacturing system designed for stability – with a focus on low price rather than resilience of supply – is confronted by a massive disruption. This is also a product that led to a resurgence of interest in more localised, closer-to-user approaches to manufacturing. Then there is a blister pack of paracetamol. This product – designed as a one-size-fits-all solution, made by the thousands of tonnes – symbolises the polar opposite of where medicine manufacturing is heading. The future is going to be based on the personal, the bespoke, the specialist, requiring whole new approaches to designing, making and using medicines. The final healthcare-related item is a single-use syringe, preloaded with blood thinner – left over from my daughter's time recovering from surgery – designed for use by patients themselves in their homes.

'You need to get rid of all this junk,' my son grumbles as he tries to make space at the table to use his laptop. I started to launch into an impassioned monologue that such linear *get rid of* thinking was out of tune with the need to move to more circular approaches, but his delivery of a rather crude phrasal verb brought that to an abrupt halt.

I stand and look down on the whole collection spread across the table. Each object provides a connection not only to one part of the manufacturing world, but to a specific period of our recent industrial history. Through these objects, I can see not only the staggering scale of progress, but also the enormity of the task remaining. Some – like the kettle controllers, the (unused) phone-battery replacement kit, an empty British Sugar paper bag – show how far we have already moved along a new path to more sustainable approaches to manufacturing and consumption. Others reflect a journey with many miles still ahead – a glass bottle of imported mineral water, that disposable face mask and the embarrassingly large number of mobile phones I have owned over the past twenty years.

I pointed out at the start of this book that, unless you are floating naked in space, you are now and pretty much at every point of your life in direct connection with and using or consuming manufactured products.

So we started a journey through the manufacturing world to mentally reconnect us to the systems that make and deliver these products. Completing that journey has also helped us to answer two really big questions:

> Why is our modern manufacturing world so fragile and so damaging to the planet? What could be done make it less fragile and damaging?

I hope you now feel you have reasonable answers to both but can also see that the task ahead remains immense.

—

Communication continues to be a key problem. Those of us who work in manufacturing need to get much better at explaining the how and why of what we do. Many manufacturing firms are really good at this, opening their doors to visitors, playing key roles within their neighbourhoods and making sure that the voice of manufacturing is present at discussions of how we grow our economies and support our communities. But we know that much more needs to be done.

Discussion of manufacturing often sits mired in unhelpful, outdated and inaccurate perceptions. Though this varies enormously country by country, the oft-heard refrain in the UK is simple: *We don't make things any more, and we don't need to. Other people can do that for us. We've got a great financial and business services sector. So why [Tim] do you keep banging on about manufacturing?* At this point, with jaw clenched and knuckles white, I calmly point them to the work of economists such as Ha-Joon Chang and the great examples of Switzerland and Singapore – the two most manufacturing-intensive but well balanced economies in the world.

At other times, I find that perceptions of our engagement with manufacturing are quite bimodal. There is the view that making things is fundamentally a 'good thing' for us – as individuals, as communities – to be engaged with.[1] But there is also the view that manufacturing is dirty, noisy, smelly and dangerous – and best kept out of sight and out of mind.

If we are to have any chance of addressing the existential challenges we face, we must redouble our efforts to engage the next generation in the importance and excitement of manufacturing, and remind them of what this book has shown: that we are all part of the manufacturing world. But as I type these words, a recent chat with a colleague comes back to mind. We were talking about the importance of schoolchildren being encouraged to visit factories. My colleague who, like me, grew up in the UK in the 1970s and 1980s said, 'Yeah, my school used to arrange trips to factories. But the pretty clear message the teachers wanted us to take away was "if you don't work hard at school, this is where you'll end up".' The message we pass on to today's future workforce must be the polar opposite.

I believe that the core of the message we need to share should also be focused on the sweet spot identified by Braungart and McDonough in *Cradle to Cradle*: framing our collective efforts to transform our manufacturing world not only in terms of *planetary survival*, but in terms of *human thriving*.

Despite the scale, uncertainty and complexity of the challenges our planet and our communities face today, there is one thing of which I am now more than ever convinced: it is manufacturing that will deliver a more sustainable, more resilient and more equitable future for us all. And despite what I said at the start of this book, manufacturing is clearly not 'another world' – it is something of which we are all a part, and in which we all play a role.

And together, we really can manufacture a better world.

I hope that by choosing to read this book, you now feel this too.

Acknowledgements

This book has had a long gestation period, and its creation has relied on kind support from a huge number of people and organisations.

Firstly, I owe a debt of thanks to all those within and connected to the community that is the University of Cambridge Institute for Manufacturing. So many current and former colleagues and students have encouraged and supported me over recent years to put down in words what we believe: manufacturing is a fundamental and fascinating part of all our lives, and is something that, if done right, can make our world a better place for all.

I would like to note special thanks to Dr John C. Taylor OBE for his generous support for all my work. The time he gave for great discussions on innovation and manufacturing during our regular meetings and calls helped define and refine the core messages of this book.

I would also like to acknowledge the role of the Sharman Fund in providing support for our institute's outreach work with schools, which allowed the messages of this book to be stress-tested with some of the toughest of audiences.

There are then all those who have been involved in the actual production of the book. I would like to thank Anna Ploszajski and Simon Hall for helping me learn the art of accessible writing, and Laura Macdougall of United Agents, who bravely took me on and found wonderful publishers, first with Faber & Faber for the UK and then ECCO/HarperCollins for the US. Thanks to commissioning editors Fred Baty and Mo Hafeez at Faber, and Sarah Murphy at ECCO, who worked with me to convert my naive scribblings into a real book. I would also like to thank Sam Wells, who provided such effective copy-editing, Jo Stimfield for her amazing project-editing skills, and Lauren Nicoll, Sophie Clarke and Jess Kim for ensuring the wider world got to hear about this book. I am very grateful to the wider teams at Faber and ECCO for all their hard work in getting this book into print, and for so patiently filling in the extensive gaps in my knowledge about how a book gets published.

I'd like to thank the Gatsby Charitable Foundation for supporting the background data collation for this project, and Carl-Magnus von Behr and Lizzy Martin who provided such responsive and effective research assistance. I would also like to acknowledge the dozens of colleagues and friends who kindly provided feedback throughout the whole project and commented on specific sections of this book. All those encouraging chats and helpful insights from people including Claire Barlow, William Barnes, Alexandra Brintrup, Charles Boulton, Abi Bush, Robin Clark, Fabio Confalone, Diane Coyle, Maharshi Dhada, Athene Donald, Steve Evans, Sarah Fell, Niamh Fox, Martha Geiger, Clare Gilmour, Rob Glew, Kevin Gotts, Maggie Harriss, John Henderson, Ward Hills, Lili Jia, Saul Jones, Mukesh Kumar, David Leal Ayala, Samantha Mayo, Duncan McFarlane, Vanessa McNiven, Dai Morgan, Dan Northam Jones, Eoin O'Sullivan, Mark Phillips, Gary Punter, Svetan Ratchev, Dave Scott, Ettore Settanni, Jag Srai, Ian Stafford-Allen, Dan Summerbell, Elizabeth Tofaris, Sarah Wightman, Alison Wright, Man Hang Yip and many, many more have helped shape this book. Particular thanks are due to Elizabeth Garnsey, Mike Gregory, John Lucas, John McManus, Tom Ridgman and Thomas Sutton for providing detailed and constructive feedback on drafts of the whole book. Any errors that remain are down to me.

So many people from so many companies, from all corners of the manufacturing world kindly let me into their factories, labs and offices, explained what they do and answered my questions so clearly. Some are named and featured in the main text, but there are many more who provided background and context. I am very grateful to each and every one of them for being so generous with their time and open with their views.

And then, my family – Nicola, Alastair and Elizabeth – who have been so wonderfully understanding and supportive while I worked on this extracurricular project over many weekends, evenings and holidays. Special thanks are due to my mother for her regular guilt-inducing asking of 'How's the book going?' and to my brothers, Pete, Matt and Luke, for so much moral support over the years.

Finally, there are two people who inspired me to write and who I hope would have been pleased to see this book in print: my father, and my friend Finbarr. Sadly, neither is around for me to hand them a copy.

Notes

Prologue

1 Climate Change Committee, 'The Sixth Carbon Budget', 2020, https://www.theccc.org.uk/wp-content/uploads/2020/12/Sector-summary-Manufacturing-and-construction.pdf.
2 Tim Harford talks about this in his book *How to Make the World Add Up*. And the definition can be found in a paper by Rozenblit and Keil (2002): 'People feel they understand complex phenomena with far greater precision, coherence and depth than they really do; they are subject to an illusion – an illusion of explanatory depth. The illusion is far stronger for explanatory knowledge than many other kinds of knowledge, such as that for facts, procedures or narratives.' Rozenblit, L. and Keil, F. (2002). 'The misunderstood limits of folk science: an illusion of explanatory depth'. *Cognitive Science*, 26:521–562. https://doi.org/10.1207/s15516709cog2605_1.
3 There's a wonderful example of this, documented by designer Gianluca Gimini. He asked people to do a simple thing: draw a picture of a bicycle, showing how it works. Sounds simple? Take a look at Velocipedia (https://www.behance.net/gallery/35437979/Velocipedia) to see how well people managed to complete this task. He then took these sketches and produced 3D CAD images of how these sketches would actually look and how dysfunctional they would be in real life. Velocipedia, https://www.behance.net/gallery/77793195/Velocipedia-IRL.
4 This approach is similar to what has been called the 'Feynman technique', named after the physicist Richard Feynman. '[. . .] in which a person attempts to write an explanation of some information in a way that a child could understand, developing original analogies where necessary. When the writer reaches an area which they are unable to comfortably explain, they go back and re-read or research the topic until they are able to do so.' Wikipedia, https://en.wikipedia.org/wiki/Learning_by_teaching.

1: Magic

1 Lora Jones, 'Coronavirus: What's behind the great toilet roll grab?', 26 March 2020, https://www.bbc.co.uk/news/business-52040532.

2 Ben Smee, 'Coronavirus: Australian newspaper prints extra pages to help out in toilet paper shortage', 5 March 2020, https://www.theguardian.com/media/2020/mar/05/australian-newspaper-prints-extra-pages-to-help-out-in-toilet-paper-shortage.

3 Ploszajski, A. (2022). 'From tree to toilet: Engineering loo roll'. *Ingenia*, 26–29. A version of this article is available at https://www.ingenia.org.uk/articles/from-tree-to-toilet-engineering-loo-roll/.

4 Li, Y., Mei, B., and Linhares-Juvenal, T. (2019). 'The economic contribution of the world's forest sector'. *Forest Policy and Economics*, 100:236–53. https://doi.org/10.1016/j.forpol.2019.01.004.

5 For coniferous trees that provide softwood, harvesting takes place at around 40 years; for broadleaved trees that provide hardwood, it can be up to 150 years. Forestry England, 'From tree to timber', https://www.forestryengland.uk/timber-uses-of-wood. Another point: the older the tree, the softer the paper. Cloud Paper, 'The Softer the Toilet Paper, the Older the Tree', 5 March 2021, https://cloudpaper.co/blogs/cloud-paper-blog/the-softer-the-tp-the-older-the-tree.

6 Nowadays, not all cellulose for toilet paper comes from chopping down trees. To reduce the amount that need to be grown and felled, around 11 per cent of toilet paper in the UK is not made from pulp from freshly felled trees but uses recycled paper. Statista, August 2023, https://www.statista.com/statistics/303006/toilet-paper-usage-by-type-in-the-uk/.

7 Anna Lambert, 'British grocers reduce SKU counts in bid to compete with discounters', 18 July 2018, https://www.newfoodmagazine.com/news/72145/uk-grocers-discounters.

8 Garbe, L., Rau, R., and Toppe, T. (2020). 'Influence of Perceived Threat of Covid-19 and HEXACO Personality Traits on Toilet Paper Stockpiling'. *PLoS ONE*, 15(6):e0234232. https://doi.org/10.1371/journal.pone.0234232.

9 Jen Wieczner, 'The case of the missing toilet paper: How the coronavirus exposed US supply chain flaws', 18 May 2020, https://fortune.com/2020/05/18/toilet-paper-sales-surge-shortage-coronavirus-pandemic-supply-chain-cpg-panic-buying/.

10 *The Economist*, 'A series of shortages threatens EU supply chains', 10 November 2022, https://www.economist.com/business/2022/11/10/a-series-of-shortages-threatens-eu-supply-chains.

11 Everett Wash, 'Boeing Celebrates Delivery of 50th 747-8', 29 May 2013, https://boeing.mediaroom.com/2013-05-29-Boeing-Celebrates-Delivery-of-50th-747-8.

12 According to Taiichi Ohno, the founding father of the Toyota Production System (the archetype of lean manufacturing), waste comes in seven flavours: transportation, inventory, motion, waiting, overprocessing, overproduction and defects. Ohno, T. (1988). *Toyota Production System: Beyond Large-Scale* Production. Productivity Press.

2: Make

1 It's a bit more than just sketching a picture – see Value Based Management, 'Summary of Value Stream Mapping', https://www.valuebasedmanagement.net/methods_value_stream_mapping.html.

2 Tim Hayward and Alison Wright's combined history book and recipe book will give you the full story and quite likely inspire you to start baking some of their recipes. See Hayward, T., and Wright, A. (2019). *Fitzbillies: Stories and recipes from a 100-year-old Cambridge bakery*. Quadrille/Hardie Grant Publishing.

3 If you haven't seen what happens in a large-scale food factory, I'd recommend a look at some of the videos I've posted on this book's website – www.timminshall.com.

4 To give that a sense of scale, the daily consumption of sugar in the UK is around five thousand tonnes (Dan Roberts, 'Sweet Brexit: what sugar tells us about Britain's future outside the EU', 27 March 2017, https://www.theguardian.com/business/2017/mar/27/brexit-sugar-beet-cane-tate-lyle-british-sugar).

5 That seems like quite a lot to me, but Hyundai's factory in Ulsan produced 1,600,000 cars in 2023. Jang Seob Yoon, 'Production volume of Hyundai Motor Company in Ulsan, South Korea from 2015 to 2023', 3 April 2024, https://www.statista.com/statistics/1176271/hyundai-motor-company-production-volume-ulsan-south-korea/.

6 *Twotogether*. (2010). 'Hainan PM 2 – die größte Papiermaschine der Welt'. *Twogether – Magazin für Papertechnik*, 16–19. https://voith.com/corp-de/voith-paper_twogether31_de.pdf.

7 *Steiermark*, 'Barrierefreies Klopapier', 26 August 2016, https://steiermark.orf.at/v2/radio/stories/2792930/index.html.

8 'Ray Dolby Centre (Cavendish Laboratory)', https://www.em.admin.cam.ac.uk/what-we-do/development-estate/building-projects/ray-dolby-centre-cavendish-iii.

9 If you want to delve more into the history of the factory, Joshua Freeman's book *Behemoth* is a great place to start. His book presents a rich map of the evolution of the modern giant factory – particularly the 'landmark' factories in the UK, US, Russia and China that punctuated the past three centuries. Freeman, J. B. (2018). *Behemoth: A history of the factory and the making of the modern world*. Norton.

10 For those wanting to know more about the Industrial Revolution, I'd recommend reading Chapter 2 of Davies, E. (2011). *Made In Britain*. Little Brown, and the front end of Freeman, J. B. (2018). *Behemoth: A history of the factory and the making of the modern world*. Norton. And the human side of this revolution is explored in Griffin, E. (2013). *Liberty's Dawn: A People's History of the Industrial Revolution*. Yale University Press.

11 For a description, see Chapter 22 in Harford, T. (2020). *The Next Fifty Things That Made the Modern Economy.* The Bridge Street Press.
12 Winchester, S. (2018). *Exactly: How Precision Engineers Created the Modern World.* William Collins. pp. 89–90.
13 Via some frankly outrageous behaviour by one Eli Whitney. See pages 94–97 in Winchester, S. (2018). *Exactly: How Precision Engineers Created the Modern World.* William Collins.
14 Quinn, T., and Kovalevsky, J. (2005). 'The development of modern metrology and its role today'. *Philosophical Transactions of the Royal Society A: Mathematical, Physical and Engineering Sciences*, 363(1834): 2307–2327. https://doi.org/10.1098/rsta.2005.1642.
15 Winchester, S. (2018). *Exactly: How Precision Engineers Created the Modern World.* William Collins. p. 60.
16 Winchester, S. (2018). *Exactly: How Precision Engineers Created the Modern World.* William Collins.
17 This is something observed with the arrival of many new technologies. See: *The Economist*, 'Solving the paradox', 23 September 2000, https://www.economist.com/special-report/2000/09/23/solving-the-paradox; and von Tunzelmann, G. N. (1978). *Steam Power and British Industrialization to 1860.* Clarendon Press.
18 Freeman, J. B. *Behemoth: A history of the factory and the making of the modern world.* Norton. p. 121.
19 There are earlier examples of such an approach. For example, there was the 'Long Shop' of the late 1800s at Leiston in Suffolk, now a museum – https://www.longshopmuseum.co.uk/.
20 Krafcik, J. F. (1988). 'Triumph of the Lean Production System'. *Sloan Management Review, 30*(1):41.

3: Move

1 Cambridge Dictionary, https://dictionary.cambridge.org/dictionary/english/logistics.
2 Felix Richter, 'China Is the World's Manufacturing Superpower', 4 May 2021, https://www.statista.com/chart/20858/top-10-countries-by-share-of-global-manufacturing-output/.
3 It wasn't always like this. For UK bike buyers back in the 70s and 80s, most bikes were UK-manufactured, with around one million per year being manufactured in the town of Nottingham alone. Robert Wright, 'Raleigh balances nostalgia with pragmatism in shift to electric', 20 May 2023, https://www.ft.com/content/c6660a89-af98-40e9-880a-295c5fa898ae.
4 Bike News, 'China Produced 76 Million Bikes and 45 Million E-bikes in 2021', 22 June 2022, https://www.bikenews.online/index.php?route=bossblog/article&blog_article_id=505#.

5 For really vivid descriptions of the type of seaborne journey my bike went on, see George, R. (2013). *Deep Sea and Foreign Going*. Portobello. or Clare, H. (2014). *Down to the Sea in Ships*. Vintage.

6 Should you wish to know how far and by what route your various products have travelled by sea, this site is quite useful: http://ports.com/sea-route/port-of-shenzhen,china/port-of-felixstowe,united-kingdom/.

7 Department for Transport, 'Port freight annual statistics 2021: Cargo information and arrivals', 27 July 2022, https://www.gov.uk/government/statistics/port-freight-annual-statistics-2021/port-freight-annual-statistics-2021-cargo-information-and-arrivals.

8 Edward Humes paints a wonderfully clear picture of the external logistics required to make such a device. Humes, E. (2016). *Door to Door: The Magnificent, Maddening, Mysterious World of Transportation*. Harper Collins.

9 Jake Hardiman, 'In Photos: How An Airbus A350 Is Built', 3 August 2021, https://simpleflying.com/airbus-a350-construction/.

10 UK Cold Chain Federation. 'Compliance', https://www.coldchainfederation.org.uk/compliance/.

11 William Dodds, 'Unilever grants industry access to patents that support ice cream storage at higher temperatures', 9 November 2023, https://www.foodmanufacture.co.uk/Article/2023/11/09/Unilever-offers-industry-access-to-ice-cream-patents.

12 A useful summary of the perceived benefits of outsourcing, written when this really took off as a popular strategy for manufacturing firms, is given in Quinn, J. B. (1999). 'Strategic Outsourcing: Leveraging Knowledge Capabilities'. *Sloan Management Review*, 4(4). At the extreme end, some car companies outsource 100 per cent of their manufacturing to others (https://www.magna.com/stories/inside-automotive/flexible-solutions/contract-manufacturer). However, others may seek to bring more manufacturing back in-house: EVannex, 'Tesla's Vertical Integration Is Something Automakers Are Eager to Copy', 20 April 2022, https://insideevs.com/news/580977/tesla-vertical-integration-automakers-copy/.

13 Levinson, M. (2016). *The Box: How the Shipping Container Made the World Smaller and the World Economy Bigger* (2nd ed.). Princeton University Press.

14 Chang, H-J. (2022). *Edible Economics: A Hungry Economist Explains the World*. Allen Lane.

15 Ibid. p. 155.

16 Joe Myers, 'The Suez Canal in numbers', 25 March 2021, https://www.weforum.org/agenda/2021/03/the-suez-canal-in-numbers/.

17 Justin Harper, 'Suez blockage is holding up $9.6bn of goods a day', 26 March 2021, https://www.bbc.co.uk/news/business-56533250.

18 I've only used the example of the *Ever Given* getting stuck in the Suez Canal to illustrate the lack of resilience in global supply chains. There are plenty of others. For example: Brexit-related shortages in the UK of drivers to take

containers to and from the ports; shortages of AdBlue (a legally required additive for diesel engines to make them less polluting) in Germany bringing fleets of trucks to a standstill; I could go on.

19 Totally Vegan Buzz, 'Fish caught off Britain shores travel 10,000 miles before reaching supermarket shelves in the UK', 23 January 2020, https://www.totallyveganbuzz.com/news/fish-caught-off-britain-shores-travel-10000-miles-before-reaching-supermarket-shelves-in-the-uk/.

20 Pisano, G., & Shi, W. (2009). 'Restoring America's Competitiveness'. *Harvard Business Review*, 87.

21 Lucy Handley, 'Firms are bringing production back home because of the Ukraine war, China's slowdown – and TikTok', 1 June 2023, https://www.cnbc.com/2023/06/01/reshoring-more-domestic-manufacturing-due-to-supply-chain-disruption.html; MakeUK, 'No Weak Links: Building Supply Chain Resilience', 7 March 2021, https://www.makeuk.org/insights/reports/no-weak-links-building-supply-chain-resilience.

22 Beautifully described in Chapter 7 of Livesey, T. (2017). *From Global to Local: The Making of Things and the End of Globalisation*. Profile Books. pp. 116–7. Livesey describes how, for some, such relocations 'conjure up images of factories being physically torn out of the ground and shunted about. Like the old man's house in the movie *Up* with the plumbing hanging out of the bottom, the image is of these factories uprooted and placed on some giant ship landing on a beach somewhere on the coast of America or Europe in some triumphant homecoming'.

23 For example, Department for Transport (2019). *Clean Maritime Plan*, UK Government; Jesse Fahnestock and Tristan Smith, 'This new strategy is paving the way for net-zero shipping', 27 October 2021, https://www.weforum.org/agenda/2021/10/net-zero-shipping-decarbonisation-new-strategy/; and the wonderful https://www.aiazero.org/.

24 'Multi issue speed report', 11 November 2019, https://seas-at-risk.org/publications/multi-issue-speed-report/.

25 Zhong, S., Lomas, C., and Worth, T. (2022). 'Understanding customers' adoption of express delivery service for last-mile delivery in the UK'. *International Journal of Logistics Research and Applications*, 25(12): 1491–508. https://doi.org/10.1080/13675567.2021.1914563; and Liz Young, 'Online Shopping's Fast-Delivery Race Is Slowing Down', 12 April 2023, https://www.wsj.com/articles/online-shoppings-fast-delivery-race-is-slowing-down-73d4c68c.

26 European Environment Agency, 'Rail and waterborne – best for low-carbon motorised transport', 24 March 2021, https://www.eea.europa.eu/publications/rail-and-waterborne-transport.

27 Department for Transport, 'Government invests £200 million to drive innovation and get more zero emission trucks on our roads', 19 October 2023, https://www.gov.uk/government/news/government-invests-200-million-to-drive-innovation-and-get-more-zero-emission-trucks-on-our-roads;

Department for Transport, 'UK confirms pledge for zero-emission HGVs by 2040 and unveils new chargepoint design', 10 November 2021, https://www.gov.uk/government/news/uk-confirms-pledge-for-zero-emission-hgvs-by-2040-and-unveils-new-chargepoint-design; Andreas Breiter, 'Powering the transition to zero-emission trucks through infrastructure', 21 April 2023, https://www.mckinsey.com/indutries/travel-logistics-and-infrastructure/our-insights/powering-the-transition-to-zero-emission-trucks-through-infrastructure.

28 Aviation Impact Accelerator, 'Explore pathways to truly sustainable flight', https://www.aiazero.org/.

29 Claire Melamed, 'Should we stop flying in organic food?', 7 September 2007, https://www.theguardian.com/environment/2007/sep/06/ethicalliving.organics.

4: Sate

1 Aaron O'Neill, 'Life expectancy (from birth) in the United Kingdom from 1765 to 2020', 4 July 2024, https://www.statista.com/statistics/1040159/life-expectancy-united-kingdom-all-time/.

2 There's a great summary of this in Chapter 3 of Perzanowski A. (2022). *The Right to Repair: Reclaiming the Things We Own*. Cambridge University Press.

3 Nils-Gerrit Wunsch, 'Cadbury products ranked by number of consumers in Great Britain from 2020 to 2022/23',13 June 2024, https://www-statista-com.cjbs.idm.oclc.org/statistics/312031/cadbury-leading-products-in-the-uk/.

4 'Order and Delivery Summary', https://www.airbus.com/en/products-services/commercial-aircraft/market/orders-and-deliveries.

5 Should you want to know more, there's a nice summary in Chapter 3 of Baudin, M., and Netland, T. (2022). *Introduction to Manufacturing: An Industrial Engineering and Management Perspective*. Routledge.

6 'Slower Growth for AR/VR Headset Shipments in 2023 but Strong Growth Forecast Through 2027, According to IDC', 21 March 2023, https://www.idc.com/getdoc.jsp?containerId=prUS50511523.

7 Wikipedia, https://en.wikipedia.org/wiki/Post-it_note.

8 There seems to be no evidence that Henry Ford actually uttered this phrase, but we *can* say that his great-grandson did report that his great-grandfather said it: 'My great-grandfather once said of the first car he ever built, "If I had asked my customers what they wanted, they would have said a faster horse." At Ford, we're going to figure out what people want before they even know it and then we are going to give it to them.' From: 2006 January 23, Transcript of Q4 2005 Ford Motor Company Earnings Conference Call, [Key Speaker: Bill Ford, Chairman & CEO, Ford Motor Company in January 2006], Congressional Quarterly Transcriptions (NewsBank Access World News) accessed via https://quoteinvestigator.com/2011/07/28/ford-faster-horse/#f+2539+1+11.

9 The impact of social media can be immediate and dramatic. A good example of this is the partnering of Dunkin' Donuts with social media influencer Charli D'Amelio. Dunkin' saw a 20 per cent overall sales boost for cold brews the day the campaign was launched, and a 45 per cent increase in cold brew sales the following day. 'Calculating Charli D'Amelio's massive sales lift for Dunkin' – StatSocial Influencer Attribution Use Case', 11 November 2020, https://www.statsocial.com/charli-damelio-dunkin-influencer-attribution. Singer and rapper Nicki Minaj posted a picture of herself on Instagram wearing pink Croc shoes. Result? A reported 4,900 per cent increase in sales for that shoe. Daisy Jordan, 'Nicki Minaj causes 4,900% spike in pink Crocs sales', 12 May 2021, https://wear-next.com/trends/nicki-minaj-pink-crocs-clogs-birkenstocks-swedish-hasbeens/. You just need to make sure that production and logistics can cope with such a spike in demand.

10 Baudin, M. and Netland, T. (2023). *Introduction to Manufacturing: An Industrial Engineering and Management Perspective*. Routledge. p. 102.

11 'Aircraft and engine order backlog hits record high as demand continue to surge in 2023 – ADS Group', 2 October 2023, https://www.themanufacturer.com/articles/aircraft-and-engine-order-backlog-hits-record-high-as-demand-continue-to-surge-in-2023-ads-group/.

12 At the end of 1977, according to the International Data Corporation, there were about 62,000 terminals in supermarkets worldwide that could scan products. Here is a great video from the BBC in 1978 explaining supermarket IBM electronic cash registers: 'Electronic supermarket checkout terminals (1978)', https://www.youtube.com/watch?v=gR46VzCAZ8I.

13 And EPOS itself is a big market, valued at $12 billion in 2021. 'Report Overview', https://www.grandviewresearch.com/industry-analysis/point-of-sale-pos-software-market.

14 Fabio Duarte, 'Amount of Data Created Daily (2024)', 13 June 2024, https://explodingtopics.com/blog/data-generated-per-day.

15 This quote has been attributed to many people. For a list, see https://quoteinvestigator.com/2019/03/23/drill/#r+22083+1+3.

5: Change

1 https://web.archive.org/web/20140401082637/http://mosi.org.uk/media/33871691/electrickettles.pdf.

2 Strix Group plc, 'Sustainability Report 2021', https://strix.com/docs/2022/strix-esg-report-2021_8d94be16b3.pdf.

3 Standage, T. (2021). *A Brief History of Motion: From the wheel to the car to what comes next*. Bloomsbury. ch. 9.

4 These included: The 1957–58 Suez crisis, the 1973–74 OPEC embargo, the 1978–79 Iranian Revolution, 1980–81 Iran–Iraq war, 1990–91 First Persian

Gulf War, 1997–98 East Asian Crisis, the 2003 Venezuelan unrest and Second Persian Gulf War (https://econweb.ucsd.edu/~jhamilto/oil_history.pdf).

5 Early attempts at making batteries that tried to overcome the limitations of lead acid ones faced some challenges. Although lithium-based batteries can store many times more power than lead-based ones, lithium in its solid state has an unfortunate habit of bursting into flames when batteries using it go through multiple charge/recharge cycles. Creating a battery that used lithium in ionic rather than metallic form solved this problem. And that's why the batteries that power nearly all our electronic devices and EVs are called lithium-ion.

6 Chris Isidore, 'Tesla is now worth more than $1 trillion', 26 October 2021, https://edition.cnn.com/2021/10/25/investing/tesla-stock-trillion-dollar-market-cap/index.html.

7 In July 2024 they were Toyota, BYD, Ferrari, Mercedes-Benz, Porsche, BMW, and VW – 'Largest automakers by market capitalization', https://companiesmarketcap.com/automakers/largest-automakers-by-market-cap/.

8 Schumpeter, J. A. (1994) [1942]. *Capitalism, Socialism and Democracy*. Routledge. pp. 82–83.

9 Martin LaMonica, 'Tesla Motors founders: Now there are five', 21 September 2009, https://www.cnet.com/culture/tesla-motors-founders-now-there-are-five/.

10 Veejay Gahir, 'Insights on Automotive Design with Veejay Gahir', https://www.linkedin.com/learning/insights-on-automotive-design-with-veejay-gahir/how-long-does-it-take-to-get-from-the-concept-phase-to-the-sales-floor?u=0.

11 Steven Grey, 'What electric vehicles mean for the future of the auto industry', 28 August 2021, https://www.tampabay.com/opinion/2021/08/28/what-electric-vehicles-mean-for-the-future-of-the-auto-industry-column/.

12 Henry Bekker, '2023 (Full Year) Global: BYD Worldwide Sales and Car Production', 2 January 2024, https://www.best-selling-cars.com/brands/2023-full-year-global-byd-worldwide-sales-and-car-production/; IEA, 'Trends in electric cars', https://www.iea.org/reports/global-ev-outlook-2024/trends-in-electric-cars.

13 Stefan Ellerbeck, 'Electric vehicle sales leapt 55% in 2022 – here's where that growth was strongest', 11 May 2023, https://www.weforum.org/agenda/2023/05/electric-vehicles-ev-sales-growth-2022/.

14 US Securities and Exchange Commission, Tesla, Inc., https://ir.tesla.com/_flysystem/s3/sec/000095017023001409/tsla-20221231-gen.pdf.

15 Wikipedia, https://en.wikipedia.org/wiki/High-definition_optical_disc_format_war.

16 There's a lovely quote about this in a pamphlet published by the Shell-Mex Company in 1930: '"Try Brown's – the chemist – a little way down on the right; he may have some". This was the sort of reply the pioneer motorist expected, and sometimes got, when he asked where he could obtain petrol,

or benzine, as it was called before the word "petrol" had been coined. In those days the fuel problem was a very real one. Supplies were few and far between, and could be obtained only from chemists and oil shops. Such service too was uncertain, and the wise motorist ordered his fuel in advance if he contemplated a trip beyond the immediate vicinity of his home. [. . .] Contrast to-day the countless service stations and petrol pumps along the highways, the giant air liners refuelling in the desert, the ships filling their bunkers with fuel-oil in every port, whilst motor cars innumerable traverse the length and breadth of the land. [. . .] Within a few years a system has been evolved by which an unceasing supply of fuel flows from the earth to the cars which use it.' Shell-Mex, *Then and Now*, 1930, p. 51.

17 Department for Transport, 'Government takes historic step towards net-zero with end of sale of new petrol and diesel cars by 2030', 18 November 2020, https://www.gov.uk/government/news/government-takes-historic-step-towards-net-zero-with-end-of-sale-of-new-petrol-and-diesel-cars-by-2030.

18 Coyle, D., and Mei, J-C. (2022). *Diagnosing the UK productivity slowdown: Which sectors matter and why?*. The Bennett Institute for Public Policy, University of Cambridge. https://www.bennettinstitute.cam.ac.uk/wp-content/uploads/2022/04/Productivity-Slowdown-in-Manufacturing-and-Information-Industries_CoyleMei.pdf.

19 MakeUK/Sage. (2023). *Digitalise to Decarbonise*. https://www.makeuk.org/insights/reports/digitalise-to-decarbonise-report.

20 Ferry, G. (2010). *A Computer Called LEO: Lyons Tea Shops and the world's first office computer*. Harper Perennial.

21 OBR, 'The transition to electric vehicles', https://obr.uk/box/the-transition-to-electric-vehicles/.

22 In 2022, sixteen billion versus eight billion. Federica Laricchia, 'Forecast number of mobile devices worldwide from 2020 to 2025 (in billions)', 10 March 2023, https://www.statista.com/statistics/245501/multiple-mobile-device-ownership-worldwide/; UN, 'World Population Prospects 2022', https://www.un.org/development/desa/pd/sites/www.un.org.development.desa.pd/files/wpp2022_summary_of_results.pdf.

6: Connect

1 Freddy Mayhew, 'UK national newspaper sales slump by two-thirds in 20 years amid digital disruption', 26 February 2020, https://pressgazette.co.uk/news/uk-national-newspaper-sales-slump-by-two-thirds-in-20-years-amid-digital-disruption/.

2 'Unveiling the 5G Auto Factory from Geely's Zeekr in China', 14 July 2023, https://telecoms.com/interview/unveiling-the-5g-auto-factory-from-geelys-zeekr-in-china/; Michael Nash, 'Zeekr VP and plant director

discusses mass volume EV production', 29 August 2023, https://www.automotivemanufacturingsolutions.com/evs/zeekr-vp-and-plant-director-discusses-mass-volume-ev-production/44568.article.

3 David Ovadia, 'A History of Pizza'. In *Bubbles in Health 2: Novelty, Health and Luxury*. (2016). Editors: Grant M. Campbell, Martin G. Scanlon, D. Leo Pyle. Elsevier.

4 Ibid. ch. 39.

5 Sloane, J., and Monaghan, T. (2003, September 1). 'The Pioneering Pizza-Delivery Chain I Started Almost Didn't Make It out of the Oven'. *CNN Money*.

6 Jacobs, S. (2014). 'Public accommodation, universal access, and technology'. *GP Solo*, 31(2): 46–49.

7 Around three-quarters of online food delivery is done via mobile apps. 'Global Online Food Delivery Market', July 2024, https://market.us/report/online-food-delivery-market/.

8 'Intelligent Engine Health Monitoring', 29 January 2019, https://www.rolls-royce.com/media/our-stories/discover/2019/intelligent-engine-health-monitoring.aspx; CIO, 'Rolls-Royce turns to digital twins to improve jet engine efficiency', 10 June 2021, https://www.cio.com/article/188765/rolls-royce-turns-to-digital-twins-to-improve-jet-engine-efficiency.html; Ingenia, 'Engine Health Management', June 2009, https://www.ingenia.org.uk/articles/engine-health-managment/.

9 Rolls-Royce estimate that using the new system has helped them to extend the time between maintenance for some engines by up to 50 per cent, thereby helping them to reduce their inventory of parts and spares, and has improved the fuel efficiency of their engines, saving over 22 million tonnes of carbon. CIO, 'Rolls-Royce turns to digital twins to improve jet engine efficiency', 10 June 2021, https://www.cio.com/article/188765/rolls-royce-turns-to-digital-twins-to-improve-jet-engine-efficiency.html

10 'Today's cars have [. . .] about 100 million lines of code; by 2030, many observers expect them to have roughly 300 million lines of software code.' McKinsey and Co. 'Cybersecurity in automotive', March 2020, https://www.mckinsey.com/~/media/McKinsey/Industries/Automotive and Assembly/Our Insights/Cybersecurity in automotive Mastering the challenge/Cybersecurity-in-automotive-Mastering-the-challenge.pdf.

11 Edge, 'The State of Informed Bewilderment', https://www.edge.org/conversation/john_naughton-the-state-of-informed-bewilderment.

12 As this technology spread more widely, it led to the rise of secret societies such as the Luddites, formed in the early 1800s in Nottingham. Their approach was 'to meet at night on the moors surrounding industrial towns, breaking into factories to destroy machinery, burn down factories, and sometimes even exchange gunfire with company guards or soldiers'. Hauge, J. (2023). *The Future of the Factory*. Oxford University Press, p. 69.

13 Lovelace's amazingly advanced work would have allowed the true capabilities of Babbage's invention to be realised . . . had it been possible to build it at that time. Swaine, Michael R. and Freiberger, Paul A., 'Analytical Engine'. *Encyclopedia Britannica*, https://www.britannica.com/technology/Analytical-Engine.
14 Standage, T. (1998). *The Victorian Internet: The Remarkable Story of the Telegraph and the Nineteenth Century's On-Line Pioneers.* Walker Publishing Company. p. 7.
15 Many variations of these optical technologies were developed such that 'by the mid-1830s, lines of telegraph towers stretched across much of western Europe, forming a sort of mechanical Internet of whirling arms and blinking shutters'. Standage, T. (1998). *The Victorian Internet: The Remarkable Story of the Telegraph and the Nineteenth Century's On-Line Pioneers.* Walker Publishing Company. p. 16.
16 British inventors Wheatstone and Cooke also developed a fully functioning telegraph system around the same time as Morse. Their system used needles on a dial to point to letters and numbers on a board. Though eventually losing out to Morse's technology, the Wheatstone and Cooke system is famous for being the first telecommunications technology used in the arrest and conviction of a murderer, one John Tawell. Wikipedia, https://en.wikipedia.org/wiki/Cooke_and_Wheatstone_telegraph.
17 This technology was so transformative that its significance in the late 1800s has been compared to that of the internet in the late 1990s. For a great summary, see Standage, T. (1998). *The Victorian Internet: The Remarkable Story of the Telegraph and the Nineteenth Century's On-Line Pioneers.* Walker Publishing Company. The roll-out of telegraph technology also required the building of a vast physical infrastructure – the network of cables strung from poles or buried under streets or laid along the seabed. How they did the undersea part of this is simply extraordinary – see Cookson, G. (2012). *The Cable.* The History Press.
18 There was some discussion as to whether Elisha Gray and Alexander Bell discovered the telephone independently. See Wikipedia, https://en.wikipedia.org/wiki/Elisha_Gray_and_Alexander_Bell_telephone_controversy.
19 This technology also enabled a move from point-to-point communication to broadcast radio transmissions. Connelly, C. (2019). *Last Train to Hilversum: A Journey in Search of the Magic of Radio.* Bloomsbury.
20 For example, in the US, manufacturing productivity effectively doubled (Doris Goodwin, 'The Way We Won: America's Economic Breakthrough During World War II', 1 October 1992, https://prospect.org/health/way-won-america-s-economic-breakthrough-world-war-ii/), allowing them to shift from making around 2,000 aeroplanes in 1939 to a staggering 96,000 in 1944 (Wrynn, D. V. *Forge of Freedom: American Aircraft Production in World War II.* Motorbooks International. pp. 4–5). Steel production went up by 100 per cent. In the decade prior to the US entering WWII, America's

shipyards launched just twenty-three ships. During the conflict, production went up to nearly one thousand ships *per year* (National Park Service, 'World War II Shipbuilding in the San Francisco Bay Area', https://www.nps.gov/articles/000/world-war-ii-shipbuilding-in-the-san-francisco-bay-area.htm).

21 This extraordinary story is beautifully told in Georgina Ferry's *A Computer Called LEO: Lyons Tea Shops and the world's first office computer*. Harper Perennial. And there is even an app you can download to have a play with a 'virtual LEO': https://www.leo1.co.uk/iOS.

22 Ferry, G. (2010). *A Computer Called LEO: Lyons Tea Shops and the world's first office computer*. Harper Perennial. pp. 107–8.

23 For a very readable history of the wonderful shenanigans of the early days of the PC industry, see Cringely, R. X. (1996). *Accidental Empires*. Penguin.

24 Jeremy Reimer, 'From Altair to iPad: 35 years of personal computer market share', 14 August 2012, https://jeremyreimer.com/uploads/notes-on-sources.txt and https://arstechnica.com/information-technology/2012/08/from-altair-to-ipad-35-years-of-personal-computer-market-share/.

25 Diebold, J. (1955). 'Automation'. *Textile Research Journal*, 25(7).

26 Available to watch at: Duncan McFarlane, 'Should we automate?', 14 February 2022, https://www.cambridgephilosophicalsociety.org/events/event/should-we-automate.

27 BBC, 'Home Computer Terminal', *Tomorrow's World*, 15 February 2013, (20 September 1967), https://www.bbc.co.uk/programmes/p0154g7b.

28 For example: Naughton, J. (2000). *A Brief History of the Future: The Origins of the Internet*. Weidenfeld and Nicolson; and Ball, J. (2020). *The System*. Bloomsbury.

29 T.S,. Kelso, 'Starlink', https://celestrak.org/NORAD/Elements/table.php?GROUP=starlink&FORMAT=csv.

30 Alejandro Jasso, 'Small Cells, Picocells, and Microcells: The Complete Guide', 6 October 2022, https://www.wilsonamplifiers.com/blog/small-cells-picocells-and-microcells-the-complete-guide.

31 Federica Laricchia, 'Number of smartphones sold to end users worldwide from 2007 to 2023', 21 May 2024, https://www.statista.com/statistics/263437/global-smartphone-sales-to-end-users-since-2007/. And the population of the world is eight billion. So that means every year, one in six people on the planet is buying a new mobile phone. There are now twice as many mobile devices in use around the world as there are people (Federica Laricchia, 'Forecast number of mobile devices worldwide from 2020 to 2025 (in billions) ', 10 March 2023, https://www.statista.com/statistics/245501/multiple-mobile-device-ownership-worldwide/).

32 8,360,000,000 subscriptions in 2022. 'World Telecommunication/ICT Indicators Database', 12 February 2024, https://www.itu.int/en/ITU-D/Statistics/Pages/publications/wtid.aspx. Figures retrieved from: World Bank Group, 'Mobile cellular subscriptions', https://data.worldbank.org/indicator/IT.CEL.SETS.

33 World Economic Forum, 'The Fourth Industrial Revolution, by Klaus Schwab', https://www.weforum.org/about/the-fourth-industrial-revolution-by-klaus-schwab.

34 Kevin Ashton, 'That "Internet of Things" Thing', 22 June 2009, https://www.rfidjournal.com/that-internet-of-things-thing.

35 Quote attributed to science fiction writer, William Gibson. https://quoteinvestigator.com/2012/01/24/future-has-arrived/.

36 Department for Business, Energy & Industrial Strategy, 'Business population estimates for the UK and regions 2022', 6 October 2022, https://www.gov.uk/government/statistics/business-population-estimates-2022/business-population-estimates-for-the-uk-and-regions-2022-statistical-release-html.

37 This includes the digital technologies that are now being used to support shop-floor workers. See, for example, T. Netland, 'Smartwatches for job task allocation in manufacturing', 8 June 2024, https://better-operations.com/2024/06/08/smartwatches-in-manufacturing/ and https://augmented-industries.de/.

38 Kabir Ahuja, et al. 'Ordering in: The rapid evolution of food delivery', 22 September 2021, https://www.mckinsey.com/industries/technology-media-and-telecommunications/our-insights/ordering-in-the-rapid-evolution-of-food-delivery.

39 Harry Wallop, '"We are democratising the right to laziness": the rise of on-demand grocery deliveries', 12 June 2021, https://www.theguardian.com/lifeandstyle/2021/jun/12/the-rise-of-on-demand-grocery-deliveries.

40 DoorDash chief operating officer Christopher Payne as reported in the *Wall Street Journal* as cited in Kabir Ahuja, et al. 'Ordering in: The rapid evolution of food delivery', 22 September 2021, https://www.mckinsey.com/industries/technology-media-and-telecommunications/our-insights/ordering-in-the-rapid-evolution-of-food-delivery.

41 Bernard Marr, '10 Amazing Real-World Examples Of How Companies Are Using ChatGPT In 2023', 30 May 2023, https://www.forbes.com/sites/bernardmarr/2023/05/30/10-amazing-real-world-examples-of-how-companies-are-using-chatgpt-in-2023/?sh=2098f5411441.

42 Noy, S., and Zhang, W. (2023). 'Experimental evidence on the productivity effects of generative artificial intelligence'. *Science*, 381:187–92. https://www.science.org/doi/10.1126/science.adh2586; Dell'Acqua, F., et al. 'Navigating the Jagged Technological Frontier: Field Experimental Evidence of the Effects of AI on Knowledge Worker Productivity and Quality'. (September 15, 2023). Harvard Business School Technology & Operations Mgt. Unit Working Paper No. 24-013, The Wharton School Research Paper. http://dx.doi.org/10.2139/ssrn.4573321; Brynjolfsson, E., Li, D., and Raymond, L. R. (2023). 'Generative AI at Work'. NBER Working Paper. 31161. https://www.nber.org/system/files/working_papers/w31161/w31161.pdf.

43 For a really engaging exploration of these issues, I'd recommend Lawrence, N. (2024). *The Atomic Human: Understanding Ourselves in the Age of AI.* Allen Lane.

44 Curtis, S. (2023). 'Undersea information sharing' *Ingenia*, September 2023. https://www.ingenia.org.uk/articles/undersea-information-sharing/.

45 Ball, J. (2020). *The System: Who owns the internet and how it owns us.* Bloomsbury Publishing.

46 Content and application providers invest over $120 billion annually in internet infrastructure – Analysys Mason, 'The Impact of Tech Companies' Network Investment on the Economics of Broadband ISPs', October 2022, https://analysysmason.com/contentassets/b891ca583e084468baa0b829ced38799/main-report---infra-investment-2022.pdf.

47 Powering the internet is estimated to have consumed eight hundred terawatt hours in 2022 (Thunder Said Energy, 'What is the energy consumption of the internet?', 20 April 2023, https://thundersaidenergy.com/2023/04/20/what-is-the-energy-consumption-of-the-internet/). Dry, windy areas where solar and wind energy can be harvested are often the same places where water is in short supply. To give a sense of scale, a new 'hyperscale' data centre requires over a million gallons of water a day – and there are around six hundred of these now in operation globally (Olivia Solon, 'Drought-stricken communities push back against data centers', 19 June 2021, https://www.nbcnews.com/tech/internet/drought-stricken-communities-push-back-against-data-centers-n1271344; Synergy Research Group, 'Microsoft, Amazon and Google Account for Over Half of Today's 600 Hyperscale Data Centers', 26 January 2021, https://www.srgresearch.com/articles/microsoft-amazon-and-google-account-for-over-half-of-todays-600-hyperscale-data-centers). Global AI demand may be accountable for 4.2–6.6 billion cubic metres of water withdrawal in 2027, which is more than half of the water used by the United Kingdom: Pengfei Li, et al., 'Making AI Less "Thirsty": Uncovering and Addressing the Secret Water Footprint of AI Models', 6 April 2023, https://arxiv.org/abs/2304.03271; CIA, 'Total Water Withdrawal', https://www.cia.gov/the-world-factbook/field/total-water-withdrawal/.

48 One aim of the original concept was that it would allow data to flow even if one or more key nodes in the network were knocked out by a nuclear attack. For a great exploration of the development of the internet, see Naughton, J. (2000). *A Brief History of the Future: The Origins of the Internet.* Weidenfeld and Nicolson.

49 In May 2017, 200,000 PCs were infected across 156 countries, including computers in sixty NHS trusts, by the WannaCry virus. The party behind WannaCry demanded $300 in Bitcoin to unlock each affected computer, with a doubling of the charge after three days, and the threat of all data being lost if payment was not received within a week. In the Department

of Health and Social Care's (DHSC) report, it says that it estimates around £20 million was lost during the attack mainly due to lost output, followed by a further £72 million from the IT support to restore data and systems. Department of Health and Social Care, 'Securing cyber resilience in health and care', 2018, https://assets.publishing.service.gov.uk/media/5bbe1250ed915d732b99254c/securing-cyber-resilience-in-health-and-care-september-2018-update.pdf.

50 OK, at the time of writing, technically the highest number of transistors in a microprocessor is 2.6 trillion, but that is for a specialised, massive chip that sits at the heart of a supercomputer designed for AI. Dean Takahashi, 'Cerebras launches new AI supercomputing processor with 2.6 trillion transistors', 20 April 2021, https://venturebeat.com/ai/cerebras-systems-launches-new-ai-supercomputing-processor-with-2-6-trillion-transistors/. And it is odd to think that, if this book is being read in the 2030s or beyond, you will be reading this and thinking, '*92 billion? Why was he so excited about such a small number?*' Then again, it'll probably all be quantum-based qubits by then, so this part of the chapter will be a quaint historical footnote.

51 Apple, 'Apple unveils M3, M3 Pro, and M3 Max, the most advanced chips for a personal computer', 30 October 2023, https://www.apple.com/uk/newsroom/2023/10/apple-unveils-m3-m3-pro-and-m3-max-the-most-advanced-chips-for-a-personal-computer/.

52 Will Knight, 'The $150 Million Machine Keeping Moore's Law Alive', 30 August 2021, https://www.wired.com/story/asml-extreme-ultraviolet-lithography-chips-moores-law/.

53 Antonio Varas, et al., 'Strengthening the Global Semiconductor Supply Chain in an Uncertain Era', April 2021, https://www.semiconductors.org/wp-content/uploads/2021/05/BCG-x-SIA-Strengthening-the-Global-Semiconductor-Value-Chain-April-2021_1.pdf.

54 'Gartner Says Worldwide Semiconductor Revenue Grew 1.1% in 2022', 17 January 2023, https://www.gartner.com/en/newsroom/press-releases/2023-01-17-gartner-says-worldwide-semiconductor-revenue-grew-one-percent-in-2022.

55 Antonio Varas, et al., 'Strengthening the Global Semiconductor Supply Chain in an Uncertain Era', April 2021, https://www.semiconductors.org/wp-content/uploads/2021/05/BCG-x-SIA-Strengthening-the-Global-Semiconductor-Value-Chain-April-2021_1.pdf. If you add up all the spending on researching and developing new products by firms in all industries around the world, roughly half of that total is spent just on designing new semiconductors.

56 'How a semiconductor factory works', https://download.intel.com/newsroom/2023/tech101/manufacturing-101-semicon/how-a-semiconductor-factory-works.pdf.

57 Antonio Varas, et al., 'Strengthening the Global Semiconductor Supply Chain in an Uncertain Era', April 2021, https://www.semiconductors.org/

wp-content/uploads/2021/05/BCG-x-SIA-Strengthening-the-Global-Semiconductor-Value-Chain-April-2021_1.pdf.

58 See amazing stories of silicon mines at: Vince Beiser, 'The Ultra-Pure, Super-Secret Sand That Makes Your Phone Possible', 7 August 2018, https://www.wired.com/story/book-excerpt-science-of-ultra-pure-silicon/; and Douglas Heaven, 'The humble mineral that transformed the world', https://www.bbc.com/future/bespoke/made-on-earth/how-the-chip-changed-everything/.

59 Douglas Heaven, 'The humble mineral that transformed the world', https://www.bbc.com/future/bespoke/made-on-earth/how-the-chip-changed-everything/. IMHO, this is definitely something you'll want to take a look at.

60 Thomas Alsop, 'Number of companies in the semiconductor ecosystem 2021, by stage', 29 June 2022, https://www.statista.com/statistics/1287789/semiconductor-companies-by-stage/.

61 Gerrit De Vynck, 'Why Nvidia is suddenly one of the most valuable companies in the world', 30 May 2023, https://www.washingtonpost.com/technology/2023/05/25/nvidia-ai-stock-gpu-chatbots/.

62 Zoe Corbyn and Ben Morris, 'Nvidia: The chip maker that became an AI superpower', 30 May 2023, https://www.bbc.co.uk/news/business-65675027.

63 Sebastian Göke, Kevin Staight, and Rutger Vrijen, 'Scaling AI in the sector that enables it: Lessons for semiconductor-device makers', 2 April 2021, https://www.mckinsey.com/industries/semiconductors/our-insights/scaling-ai-in-the-sector-that-enables-it-lessons-for-semiconductor-device-makers.

64 Cambridgeshire City Council, 'Delivery robots now available to Cambridge residents under new pilot scheme', 15 November 2022, https://www.cambridgeshire.gov.uk/news/delivery-robots-now-available-to-cambridge-residents-under-new-pilot-scheme.

7: Merge

1 Nearly $400 billion worth of medical devices are sold each year. 'Medical Devices: market data & analysis', January 2024, https://www.statista.com/study/107264/medical-devices-report/.

2 OECD, 'OECD: Health spending', 2022, https://data.oecd.org/healthres/health-spending.htm.

3 Hingorani, A. D., et al. (2019). 'Improving the odds of drug development success through human genomics: modelling study'. *Sci Rep 9*, 18911. https://doi.org/10.1038/s41598-019-54849-w.

4 You can read all about GMP and GDP at: Medicines and Health-care products Regulatory Agency and Department of Health and

Social Care, 'Good manufacturing practice and good distribution practice', 18 December 2024, https://www.gov.uk/guidance/good-manufacturing-practice-and-good-distribution-practice.
5 Ben Collins, 'Access to new medicines in the English NHS', 28 October 2020, https://www.kingsfund.org.uk/publications/access-new-medicines-english-nhs.
6 James Gallagher and Nick Triggle, 'Covid-19: Oxford-AstraZeneca vaccine approved for use in UK', 30 December 2020, https://www.bbc.co.uk/news/health-55280671.
7 CEPI, 'CEPI 2.0 and the 100 Days Mission', https://cepi.net/cepi-20-and-100-days-mission.
8 Goodier, R. (2015). 'What startups and governments can learn from Coca-Cola', *Appropriate Technology*, 42 (3):59–61, https://www.engineeringforchange.org/news/what-startups-and-governments-can-learn-from-coca-cola/. This is an example of what has been called a 'piggyback strategy' (see Savaget, P. (2023). *The Four Workarounds: How the world's scrappiest organizations tackle complex problems*. John Murray Press).
9 *The Economist*, 'BioNTech plans to make vaccines in shipping containers', 19 February 2022, https://www.economist.com/science-and-technology/biontech-plans-to-make-vaccines-in-shipping-containers/21807708.
10 BBC, 'Drugs "don't work on many people"', 8 December 2003, http://news.bbc.co.uk/1/hi/health/3299945.stm.
11 Smith, R. (2003). 'The drugs don't work'. *BMJ*, 327. https://doi.org/10.1136/bmj.327.7428.0-h.
12 NHS, 'Improving outcomes through personalised medicine: Working at the cutting edge of science to improve patients' lives', 2016, https://www.england.nhs.uk/wp-content/uploads/2016/09/improving-outcomes-personalised-medicine.pdf.
13 Srai, J. S., et al. (2015). 'Future Supply Chains Enabled by Continuous Processing', *Journal of Pharmaceutical Sciences*, 104:840–849.https://doi.org/10.1002/jps.24343
14 NHS, 'Personalised Medicine', https://www.england.nhs.uk/healthcare-science/personalisedmedicine/.
15 Erdmann, A., Rehmann-Sutter, C., and Bozzaro, C. (2021). 'Patients' and professionals' views related to ethical issues in precision medicine: a mixed research synthesis.' *BMC Med Ethics*, 22 (116). https://doi.org/10.1186/s12910-021-00682-8.
16 World Health Organization, 'Shortage of personal protective equipment endangering health workers worldwide', 3 March 2020, https://www.who.int/news/item/03-03-2020-shortage-of-personal-protective-equipment-endangering-health-workers-worldwide.
17 Ibid.
18 The Nobel Prize, 'Information for the Public', https://www.nobelprize.org/prizes/economic-sciences/2001/popular-information/.

19 There are more case studies of mid-COVID manufacturing repurposing published at: Institute for Manufacturing, 'The power of repurposing: Case studies', 16 December 2020, https://medium.com/ifm-insights/the-power-of-repurposing-case-studies-820840658dd5.
20 Charles Boulton, Tim Minshall and Jason Naselli (2020). 'The power of repurposing: How smaller manufacturers helped the UK withstand COVID-19's first wave', Institute for Manufacturing, 16 December 2020, https://medium.com/ifm-insights/the-power-of-repurposing-how-smaller-manufacturers-helped-the-uk-withstand-covid-19s-first-wave-2ebc2419dadd.
21 National Audit Office, 'Investigation into how government increased the number of ventilators available to the NHS in response to COVID-19', 30 September 2020, https://www.nao.org.uk/wp-content/uploads/2020/09/Investigation-into-how-the-Government-increased-the-number-of-ventilators-Summary.pdf.
22 Some other countries who had been wisely preparing for an event such as COVID found themselves with a surplus. As a result, the UK found itself slightly embarrassingly receiving a donation from the German army of some of their spares, as they had stockpiled so many 'just in case'. Philip Oltermann and Dan Sabbagh, 'German army donates 60 ventilators as UK scrambles for equipment', 9 April 2020, https://www.theguardian.com/world/2020/apr/09/german-army-donates-60-mobile-ventilators-uk-coronavirus-nhs.
23 Corsini, L., Dammicco, V., and Moultrie, J. (2021). 'Frugal innovation in a crisis: The digital fabrication maker response to COVID-19. R&D Management'. 51(2) :195–210. https://doi.org/10.1111/radm.12446; Corsini, L., Dammicco, V., & Moultrie, J. (2020). 'Critical Factors for Implementing Open-Source Hardware in a Crisis: Lessons Learned from the COVID-19 Pandemic'. *Journal of Open Hardware*, 4(1):8. https://doi.org/10.5334/joh.24.
24 Ulrich, B., et al. (2022). 'Potential solutions for manufacture of CAR T cells in cancer immunotherapy'. *Nature Communications*, 13:5225. https://doi.org/10.1038/s41467-022-32866-0; Sarah Nam, Jeff Smith and Guang Yang, 'Driving the next wave of innovation in CAR T-cell therapies', 13 December 2019, https://www.mckinsey.com/industries/life-sciences/our-insights/driving-the-next-wave-of-innovation-in-car-t-cell-therapies.
25 NHS, 'What is a Virtual Ward?', https://www.england.nhs.uk/virtual-wards/what-is-a-virtual-ward/.

8: Survive

1 Maslow, A. H. (1943). 'A Theory of Human Motivation'. *Psychological Review*, 50(4):370–396.
2 Panagiotopoulou, V. C., Stavropoulos, P., and Chryssolouris, G. (2022). 'A critical review on the environmental impact of manufacturing: a holistic

perspective'. *Int J Adv Manuf Technol*, 118, 603–625. https://doi.org/10.1007/s00170-021-07980-w; International Energy Agency, 'Global CO_2 emissions by sector, 2019-2022', 2 March 2023, https://www.iea.org/data-and-statistics/charts/global-co2-emissions-by-sector-2019-2022; Hannah Ritchie, Pablo Rosado and Max Roser, 'Breakdown of carbon dioxide, methane and nitrous oxide emissions by sector', January 2024, https://ourworldindata.org/emissions-by-sector.

3 Conway, E. (2023). *Material World: A Substantial Story of our Past and Future*. WH Allen.

4 'Back in the mix', *The Economist*; London Vol. 447, Iss. 9344 (Apr 29, 2023):74.

5 Global Cement and Concrete Association, 'Cement and Concrete around the World', https://gccassociation.org/concretefuture/cement-concrete-around-the-world/; Johanna Lehne and Felix Preston, 'Making Concrete Change: Innovation in Low-carbon Cement and Concrete', 13 June 2018, https://www.chathamhouse.org/2018/06/making-concrete-change-innovation-low-carbon-cement-and-concrete.

6 The global cement market size was valued at $363.67 billion in 2022 and is projected to grow from $405.99 billion in 2023 to $544.55 billion by 2030. China is the dominant producer and consumer of cement in the world. (*Fortune*, 'Cement Market Size, Share & Industry Analysis, By Type (Portland, Blended, and Others), By Application (Residential and Non-Residential), and Regional Forecast, 2024-2032', 24 June 2024, https://www.fortunebusinessinsights.com/industry-reports/cement-market-101825). Monthly cement production in China over the last twenty-five years has increased from 7,815,000 tonnes in January 1984 to 189,785,800 tonnes in November 2023 (Trading Economics, 'China Cement Production', https://tradingeconomics.com/china/cement-production). That's an increase of over 2,300 per cent.

7 Food and Agriculture Organization of the United Nations, 'Global Food Losses and Food Waste – Extent, Causes and Prevention', 2011, https://www.fao.org/sustainable-food-value-chains/library/details/en/c/266053/.

8 WWF, 'Driven to Waste: The Global Impact of Food Loss and Waste on Farms', 19 August 2021, https://www.worldwildlife.org/publications/driven-to-waste-the-global-impact-of-food-loss-and-waste-on-farms.

9 And within the 26 per cent of greenhouse gas emissions generated by things related to food manufacturing, there is huge variation between the production of different types of food. For example, beef production generates sixty times more CO_2 than the equivalent weight of peas. Hannah Ritchie, Pablo Rosado and Max Roser, 'Environmental Impacts of Food Production', 2022, https://ourworldindata.org/environmental-impacts-of-food. You might also want to read *How Bad are Bananas?* by Mike Berners-Lee, or *Food and Climate – Without the Hot Air* by S. L. Bridle.

10 Porter, Stephen D., et al. 'Avoidable food losses and associated production-phase greenhouse gas emissions arising from application of cosmetic standards to fresh fruit and vegetables in Europe and the UK', *Journal of Cleaner Production*, 201:869–878. https://doi.org/10.1016/j.jclepro.2018.08.079.
11 Food and Agriculture Organization of the United Nations, 'Global Food Losses and Food Waste', 2011, https://www.fao.org/3/i2697e/i2697e.pdf.
12 Ibid.
13 FAO, IFAD, UNICEF, WFP and WHO, 'The State of Food Security and Nutrition in the World 2019. Safeguarding against economic slowdowns and downturns', 2019, https://www.fao.org/publications/home/fao-flagship-publications/the-state-of-food-security-and-nutrition-in-the-world/2021/en.
14 Global United, 'Global Fashion Industry Statistics', https://fashionunited.com/global-fashion-industry-statistics.
15 UN Alliance for Sustainable Fashion, 'The Clothing and Textile Industry Today', https://unfashionalliance.org/.
16 'GDP by Country', https://www.worldometers.info/gdp/gdp-by-country/.
17 Ellen MacArthur Foundation, 'A new textiles economy: Redesigning fashion's future', 2017, https://www.ellenmacarthurfoundation.org/a-new-textiles-economy.
18 'Pulse of the Fashion Industry', 2017, https://www.greylockglass.com/wp-content/uploads/2021/08/Pulse-of-the-Fashion-Industry_2017.pdf.
19 Jeanologia, https://www.jeanologia.com/.
20 'WGSN x OC&C Report: Doing more with less', https://lp.wgsn.com/WGSN-OCC-Report.html.
21 Ellen MacArthur Foundation, 'A new textiles economy: Redesigning fashion's future', 2017, https://www.ellenmacarthurfoundation.org/a-new-textiles-economy.
22 'UN Alliance aims to put fashion on path to sustainability', 12 July 2018, https://unece.org/forestry/press/un-alliance-aims-put-fashion-path-sustainability.
23 Ellen MacArthur Foundation, 'A new textiles economy: Redesigning fashion's future', 2017, https://www.ellenmacarthurfoundation.org/a-new-textiles-economy.
24 MacAskill, W. (2022). *What We Owe the Future: A million-year view*. Oneworld.
25 Brundtland, G. H. (1987). *Our Common Future: report of the world commission on environment and development*, https://sustainabledevelopment.un.org/content/documents/5987our-common-future.pdf
26 Meades, J. (1996). *Double Dutch*, BBC.
27 Johanna Lehne and Felix Preston, 'Making Concrete Change: Innovation in Low-carbon Cement and Concrete', 13 June 2018, https://www.chathamhouse.org/2018/06/making-concrete-change-innovation-low-carbon-cement-and-concrete.

28 Sarah Collins, 'Cement recycling method could help solve one of the world's biggest climate challenges', 22 May 2024, https://www.cam.ac.uk/stories/cement-recycling.
29 Tkachenko, N., et al. (2023). 'Global database of cement production assets and upstream suppliers'. *Sci Data*, 10:696. https://doi.org/10.1038/s41597-023-02599-w.
30 CarbonRe, https://carbonre.com/delta-zero/.
31 For example, the Fashion Pact brings together the CEOs of major fashion companies – from H&M to Chanel – with the aim of collaborating to deliver 'a nature-positive, net-zero future for fashion' (https://www.thefashionpact.org/); the Ellen MacArthur Foundation runs multiple programmes on redesigning the future of fashion (https://www.ellenmacarthurfoundation.org/topics/fashion/overview).
32 Thomas, D. (2019). *Fashionopolis: The Price of Fast Fashion and the Future of Clothes*. Head of Zeus/Apollo.
33 P. Smith, Aug 29, 2023, Statista, 'Value of the denim jeans market worldwide from 2022 to 2030', https://www.statista.com/statistics/734419/global-denim-jeans-market-retail-sales-value/.
34 United Nations, 'UN launches drive to highlight environmental cost of staying fashionable', 25 March 2019, https://news.un.org/en/story/2019/03/1035161.
35 Maersk, '[Infographic] Denim chronicles: The journey of Jeans', 19 February 2021, https://www.maersk.com/news/articles/2021/02/19/the-journey-of-jeans.
36 Jeanologia, https://www.jeanologia.com/.
37 One of a set of techniques bundled under the heading of 'Modern Methods of Construction' - https://assets.publishing.service.gov.uk/media/631222468fa8f5423fb0c7c0/20220901-MMC-Guidance-Note.pdf.
38 Osborne Clarke, 'Are modern methods of construction in the UK more sustainable?', 27 April 2023, https://www.osborneclarke.com/insights/are-modern-methods-construction-uk-more-sustainable.
39 Hannah Ritchie, Veronika Samborska and Max Roser, 'Urbanization', February 2024, https://ourworldindata.org/urbanization.
40 M. Shahbandeh, 'Projected vertical farming market worldwide from 2022 to 2032', 12 May 2023, https://www.statista.com/statistics/487666/projection-vertical-farming-market-worldwide/.
41 Projects like the UK's FlyZero (https://www.ati.org.uk/flyzero/) and the Global Maritime Forum's 'Getting to Zero Coalition' (https://www.globalmaritimeforum.org/getting-to-zero-coalition).
42 People like Mike Berners-Lee with *How Bad are Bananas? The Carbon Footprint of Everything* and Dana Thomas with *Fashionopolis*. See also the wonderful *Without the Hot Air* series of books (starting with David MacKay's *Sustainable Energy Without the Hot Air* and now including Sarah Bridle with *Food and Climate Change Without the Hot Air* and Julian Allwood and Jonathan Cullen with *Sustainable Materials Without the Hot Air*).

43 Ellen MacArthur Foundation, 'Ellen's Story', https://www.ellenmacarthurfoundation.org/about-us/ellens-story
44 Ibid.
45 Ellen MacArthur Foundation, ' What is a circular economy?', https://www.ellenmacarthurfoundation.org/topics/circular-economy-introduction/overview.
46 Robert Adam, 'The greenest building is the one that already exists', 24 September 2019, https://www.architectsjournal.co.uk/news/opinion/the-greenest-building-is-the-one-that-already-exists.
47 Though even that is now changing – see Johanna Lehne and Felix Preston, 'Making Concrete Change: Innovation in Low-carbon Cement and Concrete', 13 June 2018, https://www.chathamhouse.org/2018/06/making-concrete-change-innovation-low-carbon-cement-and-concrete; Concretene, https://www.concretene.co.uk/; and University of Cambridge Department of Engineering, 'Cambridge engineers invent world's first zero emissions cement', 23 May 2022, http://www.eng.cam.ac.uk/news/cambridge-engineers-invent-world-s-first-zero-emissions-cement.
48 Data from the *Architects' Journal* campaign 'Retrofirst'. Will Hurst, 'Introducing RetroFirst: a new AJ campaign championing reuse in the built environment', 12 September 2019, https://www.architectsjournal.co.uk/news/introducing-retrofirst-a-new-aj-campaign-championing-reuse-in-the-built-environment.
49 Ibid.
50 'The Greenest Building: Quantifying the Environmental Value of Building Reuse', https://cdn.savingplaces.org/2023/05/24/11/14/36/697/The_Greenest_Building_Full.pdf.
51 Data from the *Architects' Journal* campaign 'Retrofirst'. Will Hurst, 'Introducing RetroFirst: a new AJ campaign championing reuse in the built environment', 12 September 2019, https://www.architectsjournal.co.uk/news/introducing-retrofirst-a-new-aj-campaign-championing-reuse-in-the-built-environment.
52 University of Cambridge (2024). '*Building Entopia: An inside look at the new headquarters of the Cambridge Institute for Sustainability Leadership with Open Cambridge*', https://www.cam.ac.uk/stories/open-cambridge-building-entopia. More information on this project can be found at https://www.cisl.cam.ac.uk/about/entopia-building.
53 Henry Ford as quoted in Perzanowski A. (2022). 'The History of Repair'. In *The Right to Repair: Reclaiming the Things We Own*. Cambridge University Press. p49–71.
54 See, for example, 'Right to Repair Regulations: Research Briefing for the House of Commons', 24 September 2021, https://researchbriefings.files.parliament.uk/documents/CBP-9302/CBP-9302.pdf. The European Commission adopted the new Circular Economy Action Plan (CEAP)

(https://eur-lex.europa.eu/legal-content/EN/TXT/?qid=1583933814386&uri=COM:2020:98:FIN) in March 2020. In March 2021, the EU introduced new right to repair standards (Roger Harrabin, 'EU brings in "right to repair" rules for appliances', 1 October 2019, https://www.bbc.co.uk/news/business-49884827). Under these new EU standards, manufacturers have to supply spare parts for certain household appliances for up to ten years, although only professional repairers will be supported by manufacturers to carry out the repairs. Currently, the standards only apply to certain household appliances (e.g. washing machines/washer-dryers, refrigerators, dishwashers and televisions). However, the EU is currently considering expanding this right to repair to more appliances, including laptops and smartphones. For a review of the issues, see Perzanowski A. (2022). *The Right to Repair: Reclaiming the Things We Own*. Cambridge University Press.

55 Perzanowski, A. (2022). *The Right to Repair: Reclaiming the Things We Own*. Cambridge University Press. p. 57.

56 London, B. (1932). *Ending the Depression Through Planned Obsolescence*. https://upload.wikimedia.org/wikipedia/commons/2/27/London_%281932%29_Ending_the_depression_through_planned_obsolescence.pdf.

57 This quote has been attributed to many people. For a list, see https://quoteinvestigator.com/2019/03/23/drill/#r+22083+1+3.

58 CoMoUK, 'Car Club Annual Report London', 2020, https://www.como.org.uk/documents/car-club-annual-report-london-2020-summary; Emily Nagler, 'Standing Still', July 2021, https://www.racfoundation.org/wp-content/uploads/standing-still-Nagler-June-2021.pdf.

59 As reported in Emily Chan, 'Is Renting Your Clothes Really More Sustainable?' 20 September, 2021, https://www.vogue.co.uk/fashion/article/is-renting-your-clothes-really-more-sustainable.

60 In the UK, emissions from manufacturing fell by 17 per cent between 2010 and 2022 – ONS, 'UK Environmental Accounts: 2023', 5 June 2023, https://www.ons.gov.uk/economy/environmentalaccounts/bulletins/ukenvironmentalaccounts/2023. In France, from 1990 to 2020, manufacturing emissions decreased by 47 per cent. 'Greenhouse Gas Emissions and Carbon Footprint', 2021, https://www.statistiques.developpement-durable.gouv.fr/media/5653/download?inline.

61 Deloitte, 'Zero in on . . . Scope 1, 2 and 3 emissions', 12 May 2021, https://www2.deloitte.com/uk/en/focus/climate-change/zero-in-on-scope-1-2-and-3-emissions.html.

62 Carbon Trust, 'An introductory guide to Scope 3 emissions', https://www.carbontrust.com/our-work-and-impact/guides-reports-and-tools/an-introductory-guide-to-scope-3-emissions; Cambridge Industrial Innovation Policy, '"No-Excuse" strategies announced for confronting Scope 3 emissions

in manufacturing and value chains', 3 January 2024, https://www.ciip.group.cam.ac.uk/reports-and-articles/no-excuse-strategies-announced-for-confronting-scope-3-emissions-in-manufacturing-and-value-chains.

63 Catapult, 'Drive to net zero at risk from "almost useless" myriad of data, UK manufacturing's strategic research group warns', 7 October 2022, https://hvm.catapult.org.uk/news/drive-to-net-zero-at-risk-from-almost-useless-myriad-of-data/.

64 Cambridge Industrial Innovation Policy, '"No-Excuse" strategies announced for confronting Scope 3 emissions in manufacturing and value chains', 3 January 2024, https://www.ciip.group.cam.ac.uk/reports-and-articles/no-excuse-strategies-announced-for-confronting-scope-3-emissions-in-manufacturing-and-value-chains.

65 Office for Product Safety and Standards and Department for Environment, Food & Rural Affairs, 'Regulations: Waste Electrical and Electronic Equipment (WEEE): Guidance for manufacturers, importers and distributors (including retailers)', 8 January 2018, https://www.gov.uk/guidance/regulations-waste-electrical-and-electronic-equipment.

66 Better Cotton, https://bettercotton.org/.

67 Machiavelli, Niccolò. (2003). *The Prince*. Translated by George Bull. Penguin Classics. (Original work published 1532).

68 World Economic Forum, 'Circular Transformation of Industries: The Role of Partnerships', 17 January 2024, https://www.weforum.org/publications/circular-transformation-of-industries-the-role-of-partnerships/.

69 Mark Riley Cardwell, 'David Attenborough: someone who believes in infinite growth is "either a madman or an economist"', 16 October 2013, https://news.mongabay.com/2013/10/david-attenborough-someone-who-believes-in-infinite-growth-is-either-a-madman-or-an-economist/.

70 Bank of England, 'What is GDP?', 10 January 2019, https://www.bankofengland.co.uk/explainers/what-is-gdp.

71 Coyle, D., and Mitra-Kahn, B. (2019). 'Making the Future Count'. Indigo Prize Essay. http://enlightenmenteconomics.com/beta/wp-content/uploads/2021/12/CoyleMitrKahnIndogo.pdf

72 Bennett Institute for Public Policy, 'Beyond GDP', 3 May 2022, https://www.bennettinstitute.cam.ac.uk/blog/beyond-gdp-impact/.

73 Braungart, M., and McDonough, W. (2009). *Cradle to Cradle: Re-making the way we make things*. Vintage. p. 36.

74 Andrew Wasley, et al. 'More than 800m Amazon trees felled in six years to meet beef demand', 2 June 2023, https://www.theguardian.com/environment/2023/jun/02/more-than-800m-amazon-trees-felled-in-six-years-to-meet-beef-demand.

75 And there are many attempts to develop more helpful measures. See, for example, Bennett Institute for Public Policy, 'Beyond GDP', 3 May 2022, https://www.bennettinstitute.cam.ac.uk/blog/beyond-gdp-impact/.

76 Vadén, T., et al. (2020). 'Decoupling for ecological sustainability: A categorisation and review of research literature', *Environmental Science & Policy*, 112:236–44. https://doi.org/10.1016/j.envsci.2020.06.016.
77 Hannah Ritchie, 'Many countries have decoupled economic growth from CO_2 emissions, even if we take offshored production into account', 1 December 2021, https://ourworldindata.org/co2-gdp-decoupling.
78 Institute for Manufacturing, 'The Regenerative Laboratory', https://www.ifm.eng.cam.ac.uk/research/industrial-sustainability/the-regenerative-laboratory/.
79 Victoria Masterson, 'What is regenerative agriculture?', 11 October 2022, https://www.weforum.org/agenda/2022/10/what-is-regenerative-agriculture/.
80 Mike Tennant, 'Sustainability and Manufacturing', October 2013, https://assets.publishing.service.gov.uk/government/uploads/system/uploads/attachment_data/file/283909/ep35-sustainability-and-manufacturing.pdf.
81 Marsh, P. (2012). *The New Industrial Revolution: Consumers, Globalization and the End of Mass Production*. Yale University Press. p. 216.
82 Jeremy Farrar and Molly Galvin, 'Major Reforms Have Been Driven by Crisis', 2022, https://issues.org/jeremy-farrar-interview-wellcome-covid/.
83 MakeUK, 'Manufacturing Fact Card 2023 Report', 2023, https://www.makeuk.org/insights/reports/manufacturing-fact-card-report-2023.
84 See, for example: Crawford, M. (2009). *The Case for Working with Your Hands: Or Why Office Work Is Bad for Us and Fixing Things Feels Good*. Penguin; and Goodhart, D. (2020). *Head, Hand, Heart: The Struggle for Dignity and Status in the 21st Century*. Penguin.
85 Braungart, M., and McDonough, W. (2009). *Cradle to Cradle: Re-making the way we make things*. Vintage. p. 11.

Epilogue

1 As well argued by Matthew Crawford in *The Case for Working with Your Hands: Or Why Office Work Is Bad for Us and Fixing Things Feels Good*. Penguin.

Index